NATIONAL ACADEMIES *Sciences Engineering Medicine*

NATIONAL ACADEMIES PRESS
Washington, DC

Exploring Key Research and Monitoring Topics for U.S. Engagement in the Fifth International Polar Year

Convened May 20–21, 2025

John Ben Soileau, *Rapporteur*

Earth Systems and Resources Program Area

Center for Health, People, and Places

Proceedings of a Workshop

NATIONAL ACADEMIES PRESS 500 Fifth Street, NW Washington, DC 20001

This activity was supported by the National Science Foundation awards 2301869 and 2232737, the National Aeronautics and Space Administration 80HQTR23FA009, and the National Academy of Sciences Maurice Ewing and Planetary Sciences Fund. Any opinions, findings, conclusions, or recommendations expressed in this publication do not necessarily reflect the views of any organization or agency that provided support for the project.

International Standard Book Number-13: 978–0-309-60212-9
Digital Object Identifier: https://doi.org/10.17226/29327

This publication is available from the National Academies Press, 500 Fifth Street, NW, Keck 360, Washington, DC 20001; (800) 624–6242; https://nap.nationalacademies.org.

The manufacturer's authorized representative in the European Union for product safety is Authorised Rep Compliance Ltd., Ground Floor, 71 Lower Baggot Street, Dublin D02 P593 Ireland; www.arccompliance.com.

Printed in the United States of America.

Suggested citation: National Academies of Sciences, Engineering, and Medicine. 2026. *Exploring Key Research and Monitoring Topics for U.S. Engagement in the Fifth International Polar Year: Proceedings of a Workshop*. Washington, DC: National Academies Press. https://doi.org/10.17226/29327.

Consensus Study Reports published by the National Academies of Sciences, Engineering, and Medicine document the evidence-based consensus on the study's statement of task by an authoring committee of experts. Reports typically include findings, conclusions, and recommendations based on information gathered by the committee and the committee's deliberations. Each report has been subjected to a rigorous and independent peer-review process and it represents the position of the National Academies on the statement of task.

Proceedings published by the National Academies of Sciences, Engineering, and Medicine chronicle the presentations and discussions at a workshop, symposium, or other event convened by the National Academies. The statements and opinions contained in proceedings are those of the participants and are not endorsed by other participants, the planning committee, or the National Academies.

Rapid Expert Consultations published by the National Academies of Sciences, Engineering, and Medicine are authored by subject-matter experts on narrowly focused topics that can be supported by a body of evidence. The discussions contained in rapid expert consultations are considered those of the authors and do not contain policy recommendations. Rapid expert consultations are reviewed by the institution before release.

For information about other products and activities of the National Academies, please visit www.nationalacademies.org/now /purpose.

WORKSHOP PLANNING COMMITEE FOR EXPLORING KEY RESEARCH AND MONITORING TOPICS FOR U.S. ENGAGEMENT IN THE FIFTH INTERNATIONAL POLAR YEAR[1]

CRAIG LEE (*Chair*), University of Washington
LILIAN ALESSA, University of Idaho
JENNY L. BAESEMAN, International Polar Consultant
SCOTT GOETZ, Northern Arizona University
JAMIN GREENBAUM, Scripps Institution of Oceanography, University of California—San Diego
MICHAEL HARTINGER, Space Science Institute, University of California—Los Angeles
CANA ULUAK ITCHUAQIYAQ, Virginia Polytechnic Institute and State University (*as of January 28, 2025*)
ANDREW J. LLOYD, Lamont-Doherty Earth Observatory, Columbia University
BRITNEY E. SCHMIDT, Cornell University

Staff

APRIL MELVIN, Associate Director
MORGAN DISBROW-MONZ, Program Officer
CAROLINE BELL, Program Officer
LINDSAY MOLLER, Research Associate
ANNIE MANVILLE, Senior Program Assistant

[1] The National Academies of Sciences, Engineering, and Medicine's planning committees are solely responsible for organizing the workshop, identifying topics, and choosing speakers. The responsibility for the published Proceedings of a Workshop rests with the workshop rapporteur.

Reviewers

This Proceedings of a Workshop on *Exploring Key Research and Monitoring Topics for U.S. Engagement in the Fifth International Polar Year* was reviewed in draft form by individuals chosen for their diverse perspectives and technical expertise. The purpose of this independent review is to provide candid and critical comments that will assist the National Academies of Sciences, Engineering, and Medicine in making each published proceedings as sound as possible and to ensure that it meets the institutional standards for quality, objectivity, evidence, and responsiveness to the charge. The review comments and draft manuscript remain confidential to protect the integrity of the process.

We thank the following individuals for their review of this proceedings:

CHARLES AMSLER, The University of Alabama at Birmingham
JACQUELINE GREBMEIER, The University of Maryland Center for Environmental Science
ANDREW LLOYD, Columbia University
MATTHEW SHUPE, The University of Colorado—Boulder

Although the reviewers listed above provided many constructive comments and suggestions, they were not asked to endorse the content of the proceedings, nor did they see the final draft before its release. The review of this proceedings was overseen by **Melinda Webster** from the University of Washington. She was responsible for making certain that an independent examination of this proceedings was carried out in accordance with standards of the National Academies and that all review comments were carefully considered. Responsibility for the final content rests entirely with the rapporteur and the National Academies.

Contents

Box and Figures

BOX

FIGURES

Acronyms and Abbreviations

AGU	American Geophysical Union
AHAB	Alaska Harmful Algal Bloom Network
AI	artificial intelligence
AOOS	Alaska Ocean Observing System
APECS	Association of Polar Early Career Scientists
cal/val	calibration and validation
CBON	community-based observing network
CH_4	methane
CO_2	carbon dioxide
ECR	early-career researcher
GI	Geophysical Institute
GIA	Glacial Isostatic Adjustment
GOOS	Global Ocean Observing System
IARPC	Interagency Arctic Research Policy Committee
IASC	International Arctic Science Committee
ICECAP	International Collaborative Exploration of the Cryosphere Through Airborne Profiling
IPY	International Polar Year
ISN	Indigenous Sentinels Network

ISRO Indian Space Research Organization
ITGC International Thwaites Glacier Collaboration

KOPRI Korea Polar Research Institute

LARISSA LARsen Ice Shelf System Antarctica
LIONESS Land-Ice/Ocean Network Exploration Using Semiautonomous Systems

MOSAiC Multidisciplinary Drifting Observatory for the Study of Arctic Climate
MPA Marine Protected Area

NASA National Aeronautics and Space Administration
NISAR NASA-ISRO Synthetic Aperture Radar
NorSOOP Norwegian Ships of Opportunity Program
NOx nitrogen oxides
NSF National Science Foundation

AOAN Alaska Ocean Acidification Network

PI principal investigator
POLENET Polar Earth Observing Network

ROOS Regional Ocean Observing System

SAR synthetic aperture radar
SCAR Scientific Committee on Antarctic Research

UAF University of Alaska—Fairbanks

VR virtual reality

1

Framing the Fifth International Polar Year

This workshop of the National Academies of Sciences, Engineering, and Medicine (National Academies) was convened May 20 and 21, 2025, to examine and discuss key research areas and operational needs that could be explored during the International Polar Year (IPY) 5, with a focus on U.S. and international researcher engagement. The workshop was intended to highlight current and potential advances in approaches and technologies in polar regions to address pressing research questions; consider existing, emerging, and anticipated capabilities that could contribute to fostering research in the polar regions; examine partnership and collaboration opportunities that can be strengthened or facilitated to benefit a wide range of groups interested in participating in IPY5 efforts; and provide a forum for early-career researchers to contribute to shaping possible focus areas for IPY5. The workshop planning committee's statement of task is in Box 1-1.

This chapter summarizes the opening welcome and chair's remarks framing the workshop, the objectives of the workshop, and a brief roadmap of the proceedings that identifies which chapter contains which session, panel, and interactive activities, and their associated themes. These proceedings were prepared by a rapporteur as an accurate account of the workshop. Statements, opinions, and ideas expressed are those of the speakers and do not necessarily reflect the views of, consensus of, or endorsement by the National Academies, sponsors, planning committee, or participants. The workshop was developed and conducted under the auspices of the National Academies Polar Research Board, in collaboration with the Ocean Studies Board and Board on Earth Systems and Resources. It was sponsored by the National Science Foundation, National Aeronautics and Space Administration, and National

BOX 1-1
Statement of Task

A planning committee of the National Academies of Sciences, Engineering, and Medicine will plan a workshop to explore ways in which the United States can contribute to achieving a transformative change in transdisciplinary polar research during the Fifth International Polar Year (IPY5). In particular, the workshop will focus on identifying critically important topics that could be explored as part of U.S. engagement in IPY5, bringing in individuals representing perspectives across a range of relevant groups and disciplines. Emphasis will be placed on engaging early career individuals in this workshop, who will be at the forefront of this research by the time IPY5 occurs in 2032–2033. The workshop participants will consider priorities and key research and monitoring themes from many different sources, including recent Antarctic and Arctic activities (e.g., the Future Directions for Southern Ocean and Antarctic Nearshore and Coastal Research report, the Fourth International Conference on Arctic Research Planning, etc.) and how they can be built upon during IPY5.

Topics to be considered during the workshop include

1. Key U.S. research themes that align with the internationally developed IPY5 guidance
2. Advances in observational, monitoring, and modeling approaches and technologies in polar regions since IPY4 (2007–2008)
3. Existing, emerging, and anticipated capabilities that could contribute to U.S. research leadership in the polar regions
4. Partnerships and collaborations that the U.S. can strengthen or facilitate to benefit a wide range of stakeholders

Academy of Sciences Maurice Ewing and Planetary Sciences Fund. The artificial intelligence (AI) platform ChatGPT was used to summarize the initial transcript of the workshop to facilitate drafting the proceedings; all material generated by an AI platform in this document was fact-checked and verified by speakers to ensure accuracy of the presented information.

Opening Remarks

Craig Lee, workshop planning committee chair, University of Washington, described the IPY as a mechanism to galvanize international activity around ambitious polar research, emphasizing projects that benefit from multination collaboration and pooled resources. He noted that the 4th International Polar Year (IPY4) (2007–2008) was recent enough that many polar researchers remember both the buildup and outcomes. IPY efforts, Lee observed, also connect the North and South Poles to global observing systems and the broader Earth system. For IPY5, he noted an opportunity to link research more directly to policy and decision-making contexts.

Lee outlined a rationale for early planning. After IPY4, many of the researchers involved felt that time had been limited to advocate and assemble resources for efforts at the desired scale. Between IPY4 and the current planning for IPY5, awareness has been growing of the importance of connecting science with societal outcomes. This workshop provided an opportunity to articulate those connections. Beginning these discussions well in advance of the envisioned IPY5 may position participants to organize, articulate compelling goals, and pursue higher resource levels. He added that IPYs build on each other and that past planning in the Southern Ocean and Arctic (e.g., Interagency Arctic Research Policy Committee) can be leveraged.

Lee said that since IPY4, methods and technologies have advanced, some pioneered at lower latitudes, and suggested wider use of them at high latitudes for both process-level science and operational observation. He also noted growing societal impacts in the Arctic, such as ecosystem change and food security concerns, that motivate closer alignment between research and community priorities.

Objectives of the Workshop

Lee emphasized that the workshop's near-term aim was to surface big scientific questions and major operational topics appropriate for an IPY-scale effort. He invited participants in-person and online to articulate ideas they hoped to see documented, including those they were thinking about independently and topics that arose during session discussions. He observed that clear, coherent inputs can strengthen arguments carried forward.

Lee encouraged discussing goals that fit the unique opportunities an IPY presents, including efforts that call for extensive collaboration and resources or could seed sustainable systems or programs with life beyond the IPY period. He also urged participants to consider societal benefits linked to the science alongside scientific capability and operational motivation.

Lee noted that the workshop planning committee had posted short presentations exploring "what the poles might look like in 2030" and invited participants to view them as a reference point during and after the meeting.[1]

Roadmap of the Workshop Proceedings

The remainder of the proceedings is structured to closely follow the workshop agenda (Appendix A). The workshop included a keynote address, Overview of the Polar Regions (Chapter 2), which covered topics such as sea-level rise and ice sheet processes, ocean–ice interactions, biodiversity, economics, and communication,

[1] To complement workshop presentations and discussions, a series of invited short videos were prepared by committee members. These videos include overviews from different areas of science that explain what the polar regions are expected to look like in 2030 and related topics for IPY5. The videos are available at https://www.youtube.com/playlist?list=PLi6VVotVxseAvm-w6fSmRPck2Vx7UGhTS.

among others. Following the keynote address, the workshop included two days of sessions with moderated panels followed by audience discussions, and each day included an interactive activities (e.g., breakout sessions). In addition to in-person comments and questions, an online chat platform was utilized for side conversations and to present questions to the panelists and the participants more broadly. The online chat discussions have been integrated into the discussion sections after each panel throughout the proceedings, and at other times, explicitly identified as originating from the online chat platform.

After the keynote address, the first day of the workshop continued with a session titled, Polar Science for Society (Chapter 3), which addressed how Arctic environmental change impacts the lives of northern residents, and how changes at both poles affect lives at lower latitudes. Session two, Key Scientific Questions and Science to Meet Multidomain Operations (Chapter 4), included a panel addressing topics such as in situ observing and remote sensing, technologies and capabilities currently available or expected to be available in 2032, and opportunities for international collaboration and coordination. The first day concluded with the session, Approaches and Capabilities to Support IPY5 (Chapter 5), which discussed how to harness new technologies, capabilities, and approaches for IPY5, and how emerging research goals for IPY5 could, in turn, inform further near-term development of these technologies and approaches. The presentations for this session were followed by a group interactive activity with a plenary report-out.

The second day of the workshop began with a recap of the first day (Chapter 6), drawing out crosscutting "umbrella" themes and focusing on linking science to societal outcomes. The remainder of the second day focused on one session, Human Capacity and Collaboration (Chapter 7), which discussed how polar science is accomplished, and highlighted many of the types of individuals and teams that enable and contribute to polar research. It also discussed specific examples of how research is conducted and the practicalities and logistics required. The session began with two panels, Strengthening Human Capacity and Strengthening Research Capacity, followed by a group interactive activity with a plenary report-out.

The final panel of the session, International Coordination and Collaboration – Moving the Science Forward for IPY5 (Chapter 8), delved into questions about how to develop internationally coordinated research questions and partnerships, and the potential roles of overarching nongovernment organizations (e.g., the World Meteorological Organization) and international committees (e.g., International Arctic Science Committee and Scientific Committee on Antarctic Research) in supporting and enhancing international collaboration and coordination in the planning and implementation of IPY5. During the final panel of the workshop, Forming the Vision for the Future: Meeting Takeaways and Reflections on Next Steps (Chapter 9), participants provided key takeaways and what they viewed as opportunities to take the research questions and move toward action and scientific advancement for IPY5.

2

Keynote Address: Overview of the Polar Regions

Richard Alley, Evan Pugh University Professor of Geoscience at Pennsylvania State University, provided a keynote address that centered the International Polar Year (IPY) 5 on people—especially students and early-career researchers (ECRs)—alongside advances in observation and modeling. He observed that hands-on polar research often builds skills that translate directly to societal applications, citing former students whose field techniques later supported aircraft safety diagnostics, disaster recovery, and engineering risk assessment. IPY activities, he suggested, can excite the public and future researchers alike. Alley addressed six themes that are detailed in this chapter.

SEA-LEVEL RISE AND ICE SHEET PROCESSES

Alley noted that large and highly visible coastal-protection proposals, for example, the "Ike Dike" for Houston, plans for New York, and initiatives around the San Francisco Bay Area, illustrate the scale of investments that hinge on credible sea-level projections; he mentioned cost estimates of billions of U.S. dollars to underscore the importance of accurate projections. He pointed to evidence that some sea-level rise estimates in major assessments may run low relative to supporting literature and historical outcomes, in part because several models simplify or omit processes that can accelerate ice loss (e.g., sub-ice-shelf melt, stress damage, calving at ice fronts). He described multimodel experiments under strong warming that span outcomes from Antarctic mass gain (lowering sea level) to multimeter rise, with model structure remaining a leading source of uncertainty (Seroussi et al., 2024). In his view, an IPY-scale effort that combined field observations, remote sensing, laboratory work, theory, and data-assimilative modeling could improve parameterizations and narrow key uncertainties. Alley argued that if even a small fraction of anticipated protection costs, such as those outlined above, are directed to research, highly valuable information could be yielded.

OCEAN–ICE INTERACTIONS, CIRCULATION, AND SEA ICE

Alley described how evolving freshwater and mass fluxes from ice sheets—the time-varying ways in which ice sheets add freshwater and ice mass to the ocean—can alter atmosphere and ocean behavior, whereas many climate models historically treat ice sheets as noninteractive "white mountains" (Alley et al. 2015). Incorporating interactive ocean–ice processes, particularly those occurring beneath ice shelves, was presented as critical for developing more credible projections. He outlined potential Southern Ocean responses to increased freshwater, such as stronger surface stratification, productivity and nutrient effects, and sea-ice redistribution, noting that tilting density layers can steer warmer waters toward grounding zones, with implications for Antarctic Ice Sheet mass balance (Bronselaer et al. 2018). Alley then shifted to the North Atlantic region, describing Greenland freshwater input and a strengthening hydrologic cycle (Bamber et al., 2018), observing that public and policy maker understanding of these circulation risks can be uneven and that better quantification of freshwater fluxes and coupled effects would be useful.

Alley also discussed the role of sea ice in abrupt regional climate variability, pointing to past episodes where winter sea ice changes coincided with large temperature swings. As seasonal navigability expands, he suggested that increasing human activities lead to greater operational demands, including forecasts across time horizons (e.g., tactical through decadal), and a greater capacity for icebreaking support and rescue. Counterintuitively, as perennial sea ice retreats, total demand for icebreaking can rise as human activity increases, including increased transits through sea ice.

BIODIVERSITY, PERMAFROST, AND THE CARBON CYCLE

Alley pointed to recent research describing observations near Churchill Canada of bears constrained on land by delayed sea ice formation and increased risks of human encounters (Stroeve et al., 2024). He connected these interactions to broader Earth system concerns: paleoclimate records and recent studies that suggest that when greenhouse gases increase in the climate system, terrestrial greenhouse gas emissions (e.g., boreal forest fires, long-smoldering permafrost) contribute to increased atmospheric greenhouse gas concentrations, which can delay sea ice formation (Canadell et al., 2021; Virkkala 2025). He suggested that reconciling model–paleoclimate differences would be valuable for projecting future change.

ECONOMICS AND VALUE OF INFORMATION

Alley discussed a commonly held viewpoint that warming yields net economic benefits in the Arctic. He noted that some economists may question whether these analyses fully account for disruption duration, losses of high-value assets, and spill-overs from neighboring regions facing stress. IPY5 research that includes a robust Arctic economic assessment, he suggested, can clarify tradeoffs. More broadly, he argued that decision-quality polar information, such as observing, modeling, and forecasting outputs about polar processes and impacts (e.g., sea-level and coastal risk information, ocean–ice interaction and freshwater flux information, sea ice services), could deliver benefits that exceed program costs.

COMMUNICATION AND WORKFORCE

Alley emphasized engagement strategies that could align with potential IPY5 activities; for example, communication could be embedded in flagship efforts and paired with brief, well-supported opportunities for early-career voices (e.g., community-run observing projects, open-data dashboards, storytellers in residence). Everyday examples can ground the conversation for general audiences. One example is how GPS on cellular phones only works because engineers apply Einstein's theory of relativity to correct the difference between how satellites' clocks tick at different rates than clocks on Earth; another example is the way fundamental research underpins many life-saving medical technologies (e.g., transcatheter aortic valve replacement). Such examples can anchor conversations in the value of science before connecting audiences to polar research. This information can be carried by diverse messengers from science, local communities, and industry; universities and philanthropy can help stabilize student participation in the near term.

DISCUSSION

In response to questions, Alley proposed a two-track strategy: implement solutions that can be adjusted as projections change and invest in research to narrow those projections, acknowledging that ice-sheet behaviors driven by fracturing and calving cannot be predicted with high precision. He also spoke to sustaining student support through institutional and philanthropic partnerships to prioritize continuity of support for students and ECRs, particularly during funding uncertainty. Finally, he supported celebrating scientific advances without implying that challenges are resolved, again underscoring the value of multiple, trusted communicators.

3

Polar Science for Society

Andrew Lloyd, planning committee member from the Lamont-Doherty Earth Observatory at Columbia University, moderated the panel in the first session, Applications of Polar Science. Lloyd stressed that this panel was about actionable science: how polar-focused research and the resulting data and understanding inform decision making and impact everyday lives. Speakers were prompted with the following questions.

- What data/information do you rely on in your position? How do you use this data/information?
- Are there any areas/issues that impact where IPY5 might offer an opportunity to fuel a significant advance in understanding or data delivery that could ease your work and/or improve the quality of the services you provide?
- How does Arctic environmental change impact the lives of northern residents, and how do changes in both poles affect lives at lower latitudes?

OPERATIONAL CAPABILITIES AND OPPORTUNITIES FROM ALASKA

Robert McCoy, University of Alaska—Fairbanks (UAF), started the panel discussion by describing the Geophysical Institute (GI) at UAF as a hub for operational polar science spanning aurora and space weather, volcano monitoring, permafrost and earthquakes, sea ice, glaciers, and extreme weather. He noted that Alaska's setting—at the intersection of the Ring of Fire and auroral oval—anchors the GI's mission and creates opportunities for research that directly links to public safety and situational awareness.

McCoy also grounded the session in real-world operating conditions by describing Alaska's geophysical setting and its importance for services and decision support. Space weather—the stream of energy and particles from the Sun that disturbs Earth's magnetic field, drives aurora, and can disrupt communications, navigation, and power systems—is a frequent wintertime feature over Alaska.[1] The state hosts 54 active volcanoes, with roughly three commonly erupting at any given time (Orr et al., 2024), with current attention to activity on Mount Spurr near Anchorage. About 90 percent of Alaska is underlain by permafrost, making thaw a crosscutting concern for infrastructure and ecosystems. Seismicity is intense—roughly an earthquake every 10 minutes. Seasonal sea ice spans about a million square miles and has long supported operations such as Navy ice camps, Ice Exercise and Sea Ice Dynamics Experiment. However, retreating sea ice is becoming more of an operational hazard for Arctic communities. Alaska's ~30,000 glaciers are also melting, paradoxically lengthening coastline (Markon et al., 2018), and Fairbanks can have very poor air quality conditions compared to the rest of the state, caused by very strong inversions. With more coastline than the entire lower 48 states, Alaska is experiencing increased wind action and fetch (the distance over open water that wind travels without obstruction), leading to higher waves and stronger coastal impacts, as sea ice declines, alongside ground subsidence where permafrost thaws—trends McCoy described as accelerating. He also noted that the Arctic is warming 3–4 times faster than the global average (Rantanen et al., 2022).

Infrastructure and Data Flows

McCoy outlined GI's role as a downlink site for polar orbiting satellites, including synthetic aperture radar. He reported 12 ground stations that receive a large fraction of National Aeronautics and Space Administration (NASA) polar orbiting data and additional streams for national mapping and geospatial agencies. GI couples those feeds with a statewide network of seismic and infrasound sensors (see Figure 3-1), rocket ranges (including suborbital launches for auroral science), and a substantial uncrewed aircraft systems program for field campaigns. GI is a Department of Defense University Affiliated Research Center with a network of seismic and infrasound sensors that detect loud sounds, like nuclear explosions, volcanoes, meteors, and other geophysical phenomena.

[1] More information about space weather prediction is available at https://www.swpc.noaa.gov/.

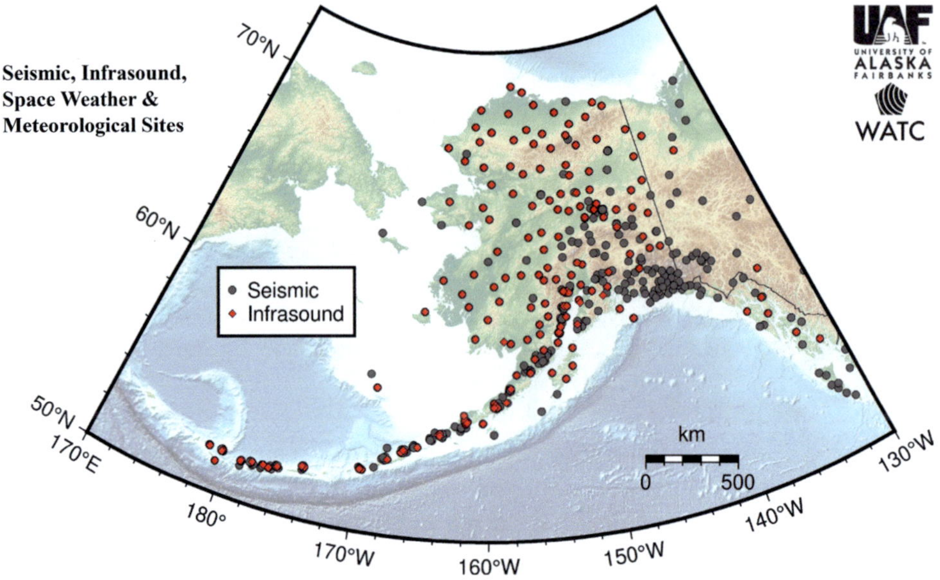

FIGURE 3-1 Seismic (gray) and infrasound (red) stations sending data in real time across Alaska.
SOURCE: Presented by Robert McCoy on May 21, 2025; created from the EarthScope Data Services archive.

Permafrost and International Collaboration

McCoy described dramatic features observed in Siberia, such as thawing permafrost releasing methane and sometimes exploding, creating crater-like blowouts. He noted that collaboration across the Bering, once routine, has paused due to the war in Ukraine and suggested that renewed international collaboration could benefit permafrost research when circumstances allow. He also described augmenting selected seismic/infrasound stations with space weather and meteorological sensors, observing that assimilating these ground observations can improve meteorological forecasts over Alaska.

Examples of Research and Opportunities in the IPY Time Frame

McCoy pointed to rapid growth in remote sensing data volumes, such as the forthcoming collaborative effort between the National Aeronautics and Space Administration (NASA) and Indian Space Research Organization, and suggested that artificial intelligence (AI) and machine learning could assist with change detection and multisensor fusion. He emphasized opportunities for uncrewed aircraft in Arctic observation, cargo delivery, and emergency response; highlighted dual use interests in sea ice monitoring for Arctic security; and mentioned interest in "science enablers," such as critical minerals and geological hydrogen, with possible roles for small modular power systems near industrial sites. He also underscored the value of combining local observations with Indigenous knowledge to coproduce useful science.

A ONE HEALTH LENS ON NORTHERN COMMUNITIES

Alisa Alexander, a clinician and director of the Center for One Health Research at UAF, framed her remarks through a One Health lens that links human, animal, and environmental health (see Figure 3-2). She noted that coordinating with community health aides and tribal organizations not only is a matter of safety and access but also aligns health, environmental monitoring, and field operations in ways that can support sustained, locally grounded observations and two-way knowledge

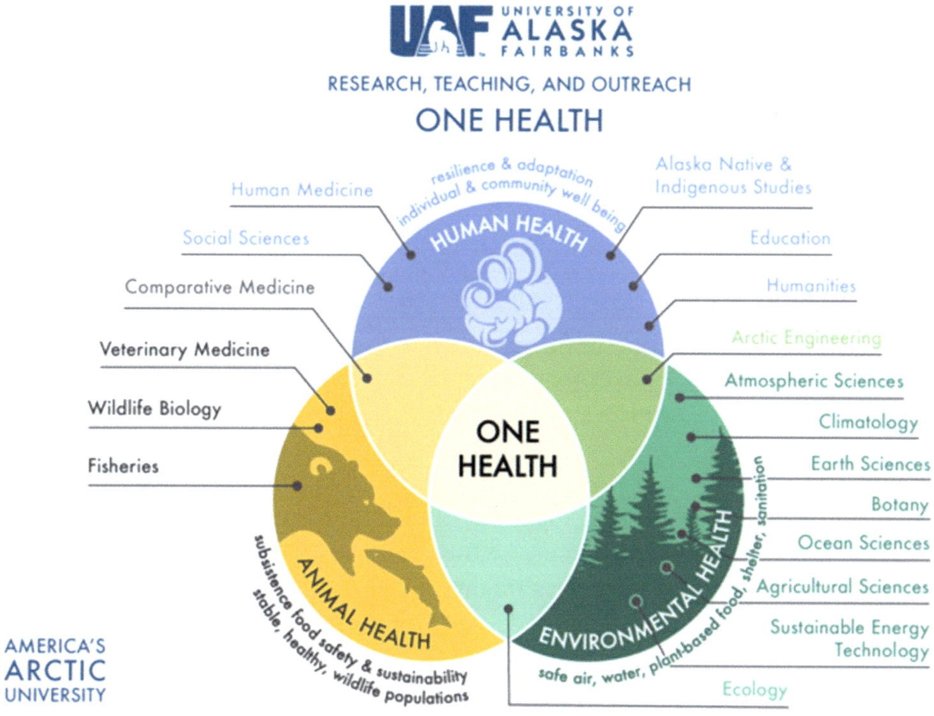

FIGURE 3-2 Venn diagram of the Center for One Health Research demonstrating the intersections of human, animal, and environmental health and the relevant departments at the University of Alaska—Fairbanks.
SOURCE: Presentation by Alisa Alexander on May 21, 2025.

exchange. She situated her perspective in remote northern communities—many of which are off the road system and rely on local clinics staffed by community health aides, with referral pathways by air. She observed that technology—such as drone delivery—can be helpful for both health care continuity and supporting researchers in remote field settings.

Health, Mobility, and Indigenous Geographies

Alexander emphasized that northern people travel widely across their territories, following traditional routes not necessarily defined by political borders, as many northerners come from nomadic heritage and have relationships with international communities. She suggested that researchers consider this mobility when planning observations, as scientific activities intersect with land use in ways that matter to residents.

Alexander described how health care is organized and delivered across rural Alaska and what that means for researchers operating in remote communities. In Alaska, tribes have compacted health services and deliver care directly; she noted that approximately 90 percent of health care statewide is provided by tribal health organizations and that village residents—Indigenous and non-Indigenous alike—receive care through their local tribal health provider.

Geography frames the care model, and most communities (Alexander said around 80 percent of communities) are off the road system, so many villages are fly-in only. Frontline care is typically from small clinics with locally trained community health aides; physicians, physician assistants, and nurse practitioners are concentrated in larger regional hubs. Referral pathways and emergency responses rely on air transport; medevac flight patterns illustrate how distance, weather, and limited infrastructure shape both routine and urgent care. Alexander observed that these conditions also apply to visiting research teams: acute medical issues can arise without warning, and evacuations for sudden conditions (e.g., appendicitis) do occur. Consequently, practical field planning can include advance coordination with local clinics and awareness of referral logistics.

Looking ahead, Alexander suggested that technology could bolster reliability and timeliness of care in remote settings. As one example, drone delivery of medications or emergency supplies might offer a backup when weather prevents flights, potentially lowering costs and stabilizing access during disruptions. More broadly, she pointed to opportunities for science–health collaborations that leverage remote sensing, communications, and logistics innovations already familiar to polar researchers, adapted to village health contexts.

Zoonoses, Permafrost, Wildfire, and Air Quality: A One Health View

Using the One Health framework, Alexander described how human, animal, and environmental health are tightly linked in northern systems but emphasized bidirectional links between the North and lower latitudes. She highlighted the annual arrival of roughly five million migratory birds as a bidirectional pathway for pathogens and contaminants between high and lower latitudes, noting avian influenza as one example. In parallel, she described how thawing permafrost, including yedoma (an understudied form of permafrost), is a potential source of mercury and other materials in waterways, and she connected these dynamics to broader ecosystem change, including prolonged salmon fishery closures in parts of Alaska with ecological, dietary, and cultural implications for northern communities.

Alexander observed that wildfire emissions datasets for northern regions are sparse and often out of date, limiting the ability to quantify carbon fluxes and assess health impacts. She referenced an analysis of the 2023 Yellowknife fires (Sutton et al., 2024) suggesting that roughly half of measured atmospheric arsenic during that episode originated from the fires, underscoring air quality concerns and the possibility that fire removes insulating soil organic matter in ways that accelerate permafrost thaw and ground instability. In her view, more current, integrated observations—linking local observing networks, remote sensing, and health surveillance—can clarify these coupled risks for communities and decision makers in the North.

Community Observing and Coproduction

Alexander described local observing networks (e.g., UAF-affiliated efforts, such as the Alaska Arctic Observatory & Knowledge Hub) as platforms for sustained observations and coproduction, integrating traditional knowledge with instrumented measurements and historical perspectives. She noted that consistent engagement can extend the temporal coverage of datasets and reveal questions that may not be obvious to visiting research teams—for example, community interest in landscape features visible from space, such as ancient caribou fences. She also referenced recent U.S. interagency planning (e.g., International Arctic Research Policy Committee [IARPC] processes) that has elevated Indigenous involvement.

LINKING POLAR RESEARCH TO SOCIETAL CHALLENGES

Jeremy Wilkinson, British Antarctic Survey, discussed how funding structures shape the engagement of polar science with societal challenges. From a U.K. perspective, he observed that agency mandates (e.g., Natural Environment Research Council, Engineering and Physical Sciences Research Council, Economic and

Social Research Council), make cross-sector and interdisciplinary proposals and projects (e.g., spanning medicine and public health, engineering, physical and social sciences) more challenging to develop and evaluate. Eligibility rules that favor permanent academic posts can also exclude important partners (e.g., Indigenous organizations, local authorities, industry operators, nongovernmental organizations, boundary organizations) that are central to turning knowledge into services and decisions.

Wilkinson contrasted the U.K. model with EU frameworks (e.g., Horizon Europe) that issue calls for research explicitly framed around societal challenges and expect multicountry, multisector consortia from the outset, including Indigenous partners and industry. However, he noted that even in the EU model, dedicated mechanisms to deliver results directly to ministries and operational agencies can be uneven, which makes proactive outreach, translation products, and knowledge-broker roles even more essential to a project's design and budget.

Implications for IPY5

Wilkinson suggested that a global initiative, such as IPY5, could convene once-in-a-generation efforts—beyond individual grants—by defining challenges clearly, supporting teams that span sectors, and engaging policy makers early so that results are positioned for use. Mission-oriented efforts could be organized around well-defined societal challenges (e.g., credible sea-level projections for coastal planning; sustained pan-Arctic fire–permafrost assessments for health and infrastructure; polar services that support navigation, fisheries, and emergency response). He also pointed to the opportunity to unify polar observing across nations to reduce duplication and close spatial/temporal gaps via shared platforms, open standards, and lead-agency/co-funding agreements and to keep data open and accessible.

DISCUSSION

Outreach and Misinformation

Oscar Schofield, Rutgers University, asked how outreach could be organized as a community rather than as many individual efforts, especially in new media and with audiences skeptical of Earth science. Wilkinson suggested partnering beyond academia, such as with artists and communicators, and across sectors, so that complex topics reach varied audiences through trusted messengers. McCoy described ongoing public engagement at UAF (e.g., public lectures and open houses, such as at the High-Frequency Active Auroral Research Program) and the value of persistent efforts in combating the scale and continued risk of misinformation. Alexander noted capacity concerns in Indigenous communities when engagement requests surge and suggested that professional science communicators can play important roles.

Sustainable Funding Models

Twila Moon, National Snow and Ice Data Center (Cooperative Institute for Research in Environmental Sciences, University of Colorado—Boulder), asked about funding models that sustain collaborations over decades rather than short cycles. Alexander described the difficulty of maintaining long-term projects beyond 3–5 year grant horizons, and McCoy added that long-term trust with communities takes time, and international cooperation is intermittent; Wilkinson pointed to durable networks in meteorology and oceanography (e.g., Argo) as examples and suggested that an IPY5 framework could explore mechanisms that extend project lifetimes.

Hazel Shapiro, IARPC Secretariat, asked Alexander whether models from tribal health care compacts could inform research funding so that communities could direct research for high-impact and locally relevant topics. Alexander called it an interesting idea and referenced Herrmann (2024), showing all the responsibilities tribes already carry. Alexander emphasized consulting tribes and Indigenous peoples regarding interest and capacity. Wilkinson underscored that the best way forward is bringing communities together in an inclusive and equitable way.

Communicating the Value of Data

An audience member asked how the importance of satellite and field data could be communicated to and recognized by the public. McCoy suggested that AI can help extract salient insights across large datasets for clearer communication. Wilkinson pointed to community-centered science projects that make satellite products relevant and tangible for local users. Alexander proposed inviting community members to ask questions that satellites might be able to address, such as "We're using this satellite to look at X—do you have questions you want us to look at?"—and referenced cultural features (e.g., caribou fences) visible from space as mutually interesting targets. Stacy Lucenson pointed out that IPY5 planning could include Indigenous peoples in both the Arctic and Antarctic, including Māori for the Southern Ocean.

Environmental Services for IPY5

Planning committee chair, Craig Lee, asked which environmental services IPY5 might support and which sectors and stakeholders could be engaged. Wilkinson mentioned observing networks, noting that multiple nations do observing, but not in a unified way, which creates both gaps and duplication: "If we could bring order to that chaos, unify how we observe the Arctic and Antarctic, and make the data open and accessible—that would be a real benefit!" McCoy pointed to themes that connect sea ice, permafrost, and satellite datasets, noting the World Meteorological Organization system as a model of international collaboration. Alexander emphasized the role of Indigenous peoples as key stakeholders and described traditional practices (e.g., cultural burning) as topics for study, with implications for forests, permafrost, air quality, and carbon cycling.

4

Key Scientific Questions and Science to Meet Multidomain Operations

The Session 2 panel, Key Scientific Questions and Science to Meet for Multidomain Operations, was moderated by Scott Goetz, planning committee member from Northern Arizona University. He framed the session to focus on exploring ambitious research goals that could leverage the span and scope of International Polar Year (IPY) 5 to achieve transformative advancements in understanding of the polar regions, their role in the broader Earth system, and the impacts of polar environmental change on society. Speakers provided examples of research areas and themes where a large effort assembled for IPY5 could facilitate transformative advances and fulfill operational needs. They were prompted to address the following questions:

- What research areas could meet multidomain operational needs (i.e., have multiple uses across research disciplines and/or applications)?
- What technologies and capabilities are currently available, or expected to be available in 2032, to enable addressing these questions? What opportunities might exist for international collaboration and coordination to facilitate addressing these questions?
- Are there other specific needs/foci within these broad examples that should be brought to the forefront of our thinking?
- What other key topics could be explored during IPY5, and what operational needs might they address?

ICE SHEETS, SEA LEVEL, AND GROUNDING ZONES

The first panelist, Jacqueline Austermann, Lamont-Doherty Earth Observatory, Columbia University, framed a central question for cryosphere–coast research: "how fast will ice sheets melt?" She described contrasts between the Greenland Ice Sheet ($\approx$7 m sea-level equivalent; lower latitude, surface melt dominant) and Antarctic Ice Sheet (West $\approx$5 m, East $\approx$55 m sea-level potential; mass loss focused at grounding zones where ice and ocean interact over bedrock in large parts below sea level). She observed that uncertainties cluster at grounding zones, where atmosphere and ocean, sub-ice-shelf processes, ice dynamics, erosion and sedimentation processes, glacial isostatic adjustment, and subglacial hydrology intersect and interact.

Because grounding zones are the regions that are most vulnerable to ice loss and where uncertainty is greatest, Austermann suggested three complementary directions:

1. **Instrumenting and modeling the grounding zone.** Integrated packages—GPS, radar, thermometry—on both grounded and floating ice, coupled with models and maintained through time, can capture accelerations and process coupling (see Figure 4-1).
2. **Interrogating paleoclimate analogs.** Records of past ice sheet retreat (e.g., last interglacial) can reveal collapse conditions.
3. **Modeling local coastal impacts.** Relative sea level varies by location; Greenland's ice sheet melt, for example, can lower sea level locally. Augmented tide gauge coverage (noted as sparse around Greenland) and bathymetric mapping can support community engagement on hazards.

GEOSPACE AND SPACE WEATHER: WHY THE POLES MATTER

Allison Jaynes, University of Iowa, explained that the poles are the main gateways through which solar-wind energy enters near-Earth space. Guided along magnetic field lines into high latitudes, it produces auroras and disturbs the upper atmosphere. She linked this to real-world impacts often concentrated at high latitudes but with global reach: communications and navigation disruptions (GPS lock loss and positioning errors), radio blackouts, satellite anomalies or failures, transpolar flight reroutes, and geomagnetically induced currents that affect power grids.

Observations, Shared Networks, and Atmospheric Links

Jaynes pointed to overlapping observing modes relevant to geospace and atmospheric communities—Global Navigation Satellite System, riometers, auroral

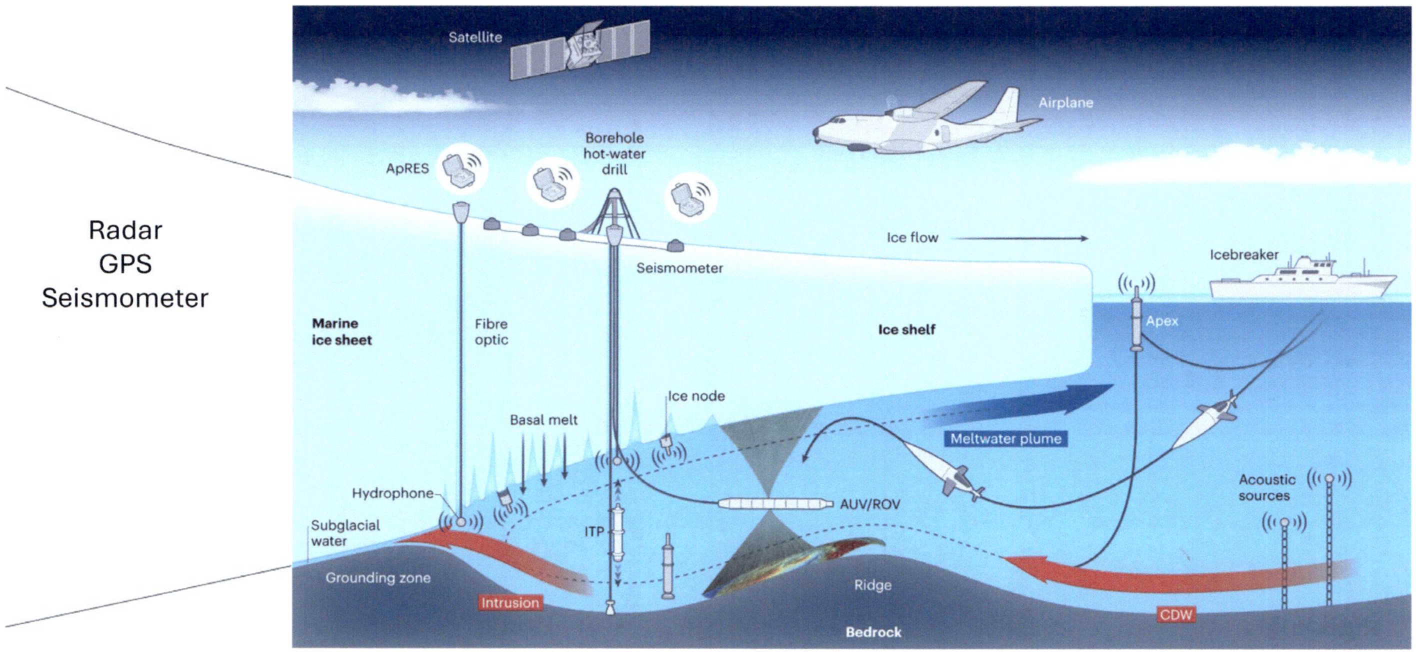

FIGURE 4-1 Range of observing systems that can be used to study the ice sheet, ocean, and bed.
SOURCE: Presentation by Jacqueline Austermann on May 21, 2025 (slightly adapted version of Figure 1 from Rignot [2023], reproduced with permission Springer Nature).

imagers, ionosondes, and radars—and noted that shared logistics (e.g., coordinated flights, joint field campaigns, international partnerships) can amplify value. Jaynes explained that energetic particles from the Sun produce nitrogen oxides (NOx) and hydrogen oxides high in the atmosphere; in polar winter, NOx can descend into the stratosphere and contribute to ozone loss, highlighting the value of whole-atmosphere models linked to polar observations.

PERMAFROST CARBON

Arctic Futures and Permafrost as a Triple Interaction

Ted Schuur, Northern Arizona University, offered a forward look—framing decisions today in terms of the world lived in by future generations. He emphasized permafrost thaw as an interaction among temperature, carbon, and ice, with consequences for infrastructure, ecosystems, and the global carbon cycle. He noted how exposure of ice-rich permafrost called "yedoma," which Alexander had mentioned in the previous session, illustrates structure and vulnerability. Drawing on a recent synthesis, Schuur described an "Arctic of the future" in the year 2087 as a summer ice-free Arctic Ocean serving as a significant shipping corridor, widespread surface–permafrost degradation, and a Greenland Ice Sheet that persists while contributing roughly one-third of the global sea-level rise (Stroeve et al., 2025).

Observing Networks and Measuring Fluxes

Maps of greenhouse gas flux sites (e.g., Figure 4-2) and measurement locations of permafrost temperature and active-layer depth appear dense at first glance but include many single-campaign locations; long-operating sites are fewer. He observed that regional to global assessments lean on a few independently run sites linked primarily by collaboration. Direct measurements of subsurface carbon processes complement remote sensing; together, they can function like a "storm warning system" for permafrost carbon, informing both global assessments and local adaptation. Schuur underscored opportunities to pair ground measurements with satellite synthesis, pursue breakthrough methods suited to rapid change, and accelerate progress through open-science practices.

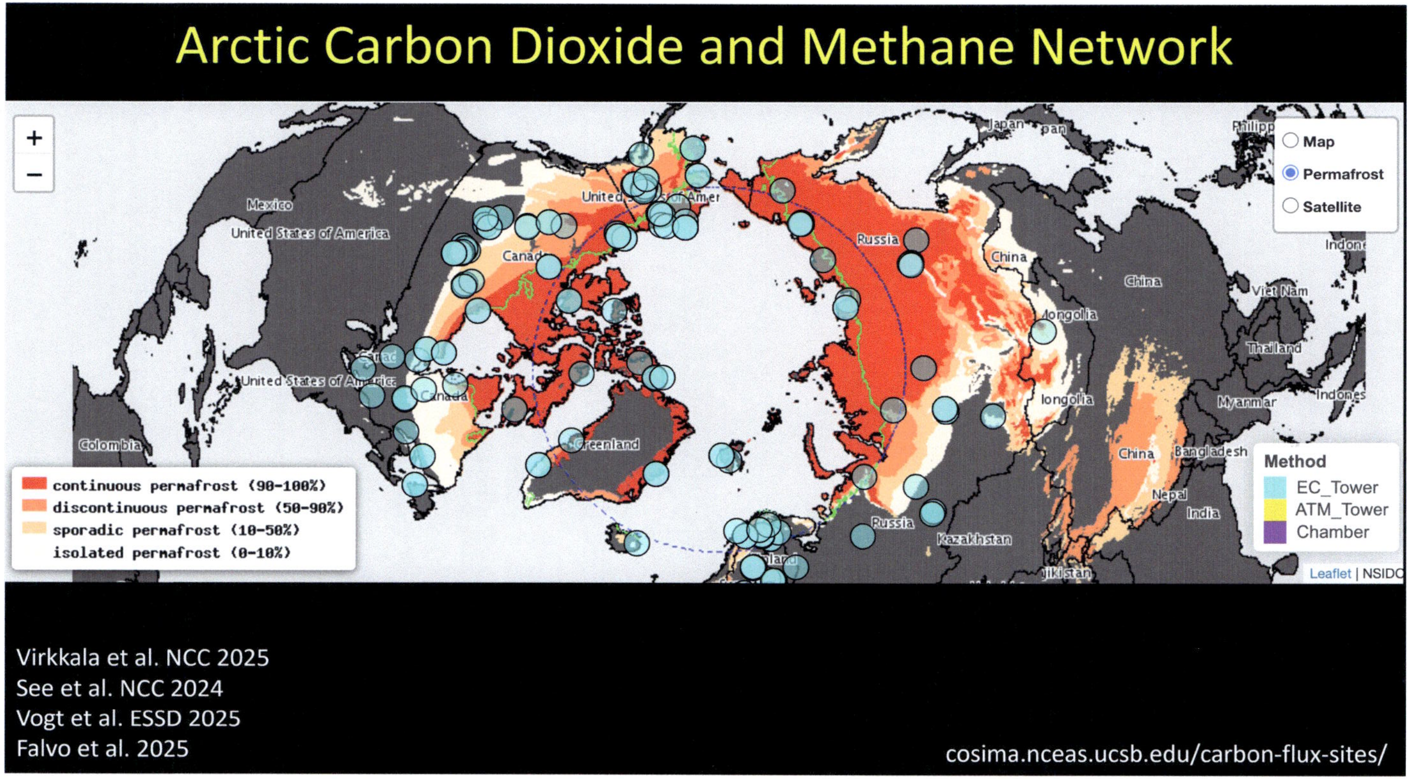

FIGURE 4-2 Arctic Carbon Dioxide and Methane Network.
SOURCE: Presentation by Ted Schuur on May 21, 2025; Vogt (2025); see also, Virkkala et al. (2025), See et al. (2024), and Falvo et al. (2025).

Partnerships, Programs, and Data Portals

Alaska Ocean Observing System (AOOS): Convener and Connector

Sheyna Wisdom, executive director of AOOS, provided an overview of AOOS as one of 11 National Oceanic and Atmospheric Administration–funded regional associations across the United States. She described how it is both a funder and a convener whose mission is to increase observing and forecasting in all regions of Alaska. Its board includes representatives from federal and state agencies, academic institutions, research facilities, industry, and Indigenous organizations. Wisdom described AOOS work as including sustained observations (e.g., ecosystem moorings, gliders, high-frequency radars, ship surveys), pilots of innovative technologies (e.g., low water level sensors, HydroBall bathymetry, image flow cytobots), and user-facing tools (e.g., portals, dashboards) codesigned with, for example, fishermen, emergency responders, industries, and researchers.

Community-Embedded Observing and Data Access

Wisdom described the Backyard Buoys project—whaling captains deploying wave buoys across the Arctic—and collaboration networks, such as the Alaska Harmful Algal Bloom Network (AHAB) and Alaska Ocean Acidification (OA) Network. Wisdom highlighted the AOOS Ocean Data Explorer (≈3,000 data layers) and a mobile app supporting mariner decisions, noting ongoing work to unite data streams and to archive hard-to-access datasets. She emphasized partnerships with Indigenous communities across multiple AOOS programs (e.g., Sea Ice for Walrus Outlook, Arctic Watch, Skipper Science, the Arctic Observing Knowledge Hub), with a focus on actionable information.

Implications for IPY5

In the context of IPY5, AOOS can use its convening function, data infrastructure, and gap-focused funding approach to support coproduced services and interoperable observing that plug into broader Arctic and Antarctic efforts, including the following:

- **Connector/convener:** Align Alaska's priorities with national and international observing plans, drawing on existing networks (e.g., OA, AHABs, Alaska Water Level Watch) and ties to other regional associations (e.g., Northeastern Regional Association of Coastal Ocean Observing Systems, Mid-Atlantic Regional Association Coastal Ocean Observing System).

- **Data infrastructure**: Use a portal as a centralized hub that links datasets, even those not in it, to streamline access and reduce redundancy for communities and end users.
- **Targeted investments:** Leverage partner funds to fill observing gaps and pilot tools that scale (e.g., wave buoys deployed by local users, community dashboards).

MARINE ECOSYSTEMS AND DISTRIBUTED BIOLOGICAL OBSERVATORIES

Jacqueline Grebmeier, University of Maryland Center for Environmental Science, discussed the rationale for sustained ecosystem observing to understand ongoing Arctic marine ecosystem change—sea ice loss, warming waters, and coastal erosion—and links to the carbon cycle. She pointed to biodiversity shifts, northward species movements, food web reorganization, community food security, fisheries migration, harmful algal blooms, and ocean acidification as issues that motivate time series observations. She also noted that large marine vertebrates reflect habitat and prey changes that can be observed through range shifts and changes in growth rates.

Distributed Biological Observatory Model and Next Steps

The Distributed Biological Observatory, initiated in the Pacific Arctic after IPY4, now offers a pan-Arctic template: standardized, repeated measurements coordinated across national efforts via networks and steering committees, enabling comparisons across basins. Grebmeier underscored integration across physics, chemistry, and biology; process studies; and new technologies (e.g., biogeochemical sensors, eDNA, passive acoustics, autonomous systems) feeding real-time, open data streams. She noted that forecasting marine ecosystem change remains challenging relative to physical predictions, but is a worthy goal for future activities. In addition, she encouraged inclusion of early-career scientists and coastal community members into planning and field activities.

EMPEROR PENGUINS AND SATELLITE TECHNOLOGY

Satellites, Animal Tags, and Assessing the Ross Sea Marine Protected Area (MPA)

Michelle LaRue, University of Canterbury, provided a prerecorded presentation. She described how advances in satellite imaging and animal-borne sensors are expanding what can be learned about Southern Ocean ecosystems, using emperor penguins to frame a larger management question: "Is the Ross Sea MPA functioning as intended?"

LaRue traced the arc from early "penguins-from-space" work—submeter optical imagery provided via U.S. resources (e.g., through National Geospatial-Intelligence Agency to National Science Foundation–funded researchers) that revealed unknown penguin colonies and supported a first global estimate—to a decade-scale reanalysis indicating ~225,000 adult birds (2018) and 66 known colony sites (LaRue et al., 2024). She noted, however, that springtime optical images may track fledgling abundance more than breeding pairs, underscoring method limitations and the value of complementary approaches. New, high-resolution synthetic aperture radar (SAR) now makes winter detection feasible, opening the door to the first wintertime estimates of breeding populations. Satellite tags deployed on 10 emperor penguins in November 2024 added behavioral and habitat context (e.g., foraging locations, dive depths, molt sites); once shed, several tags continued to drift with pack ice, effectively becoming sea-ice trackers and yielding fine-scale physical data. Taken together—optical imagery, SAR, and telemetry—these tools can map most of the species' annual cycle and habitat use, providing inputs relevant to MPA evaluation and broader ecosystem assessment. LaRue suggested that answering "Is the MPA working?" will hinge on access to the right technology, sustained investments, and cross-national, interdisciplinary collaboration—an approach that aligns well with an IPY5 portfolio leveraging U.S. satellite resources alongside partner capabilities.

DISCUSSION

Addressing Observational Gaps in Models

An audience member inquired about the biggest observational gap limiting estimates of how fast ice melts or permafrost thaws, from a modeler's perspective, and another asked which observed processes or behaviors remain poorly represented in models. Austermann responded that grounding zone observations—including GPS, radar, and thermometry integrated across grounded and floating ice—are critical; long-term GPS on the solid Earth side was also noted. Grebmeier responded that ecosystem models benefit from rate measurements (e.g., phenology, growth), especially in shelf regions that connect physical drivers to biology. Additionally, she stressed the importance of data from high-intensity areas, like ice shelves, to link physical drivers to ecological change. She said that biological phenology (e.g., blooms, migrations, reproduction) is shifting; ecological models lag physical models. She added that atmospheric profiles (e.g., radiosondes) are often sparse during Arctic field campaigns even though they can clarify links.

Schuur replied that permafrost ground ice distribution, which is heterogeneous and difficult to observe, underpins thaw, subsidence, and carbon release. Subsurface prediction remains a significant gap that IPY5 could target. Additionally, abrupt thaw dominates some landscapes, yet many models use gradual thaw; coupling to mapped ground ice could aid parameterization.

Several participants in the online chat noted the lack of representation of key atmospheric challenges, such as clouds, aerosols, precipitation, radiation, and energy budgets, among others. Additionally, an online participant noted the need for distributed measurements of atmospheric and oceanic variables, and to ensure the associated science and prediction questions for such variables can inform gaps and potentials for new observing and modeling systems for IPY5 in both polar oceans.

Revisiting and Reusing Datasets for IPY5

A participant inquired about datasets from IPY4 or the International Geophysical Year that might merit remeasurement for an IPY5 project. Schuur said that repeating datasets would be valuable; several programs are ending, but long-term records remain important. Grebmeier replied that ecological time series now reach 30–40 years in places; maintaining and expanding pan-Arctic coverage can enhance insight, with some cryosphere measurements also informing planetary analogs. Jaynes replied that the north magnetic pole has shifted substantially since IPY4, repeating observations that now is timely for both geospace and ecological questions.

5

Approaches and Capabilities to Support IPY5

Britney Schmidt, planning committee member from Cornell University, moderated Session 3 by noting how the remote, harsh environment of the polar regions presents unique challenges for observational research, complicating the task of collecting sustained measurements of the necessary spatial and temporal scales and scope. She described how polar research often requires highly specialized equipment and infrastructure (e.g., ice-strengthened vessels, ice core drilling rigs) built to withstand extremely cold conditions. Additionally, extensive, costly logistical capabilities are used to deploy and sustain researchers and equipment in remote locations. Though largely spared the environmental challenges of in situ observing, remote sensing and numerical modeling rely on in situ measurements and are thus constrained by sparse data availability; temporal and spatial resolution challenges; and other issues, such as frequent cloud cover and seasonal darkness interfering with satellite-based sensing.

Schmidt continued by noting the rapid changes in technologies and approaches that have occurred since IPY4, with advancements continuing to emerge that may enable new and expanded efforts in field, remote, and modeling research for IPY5. The session included ideas on how to harness new technologies, capabilities, and approaches for IPY5 and how emerging research goals for IPY5 could inform their near-term development. The panelists were prompted with the following discussion questions:

- What new approaches, capabilities, and technologies are available (or are anticipated to be available by 2032) that could help facilitate research advancements during IPY5?
- Are there opportunities for new interdisciplinary approaches or synergistic use of capabilities/technologies among disciplines that could be leveraged?
- Are there other capabilities or technologies that do not yet exist that are fundamental to addressing any of the priority research areas discussed? Is there a path to reaching these capabilities/technologies?

REMOTE SENSING, ACCESS, AND POWER

Matthew Siegfried, Colorado School of Mines, organized his remarks around three themes: remote sensing continuity, access to hazardous or critical interfaces, and power systems. Siegfried portrayed current capabilities as "knife edge": the National Aeronautics and Space Administration (NASA) Ice, Cloud, and Land Elevation Satellite 2 retains funding yet faces questions; NASA- Indian Space Research Organization (ISRO) Synthetic Aperture Radar (NISAR) had not launched; and future missions, such as Earth Dynamics Geodetic Explorer, were uncertain. He argued that lidar could be pivotal for snowfall and surface mass balance retrievals and NISAR for persistent, continent-wide synthetic aperture radar (SAR) coverage to estimate ice velocity; he suggested that engaging with commercial providers might bridge gaps if public missions slip.

Siegfried emphasized hazardous but consequential zones—shear margins, grounding lines, and the ice-bed interface—where autonomous systems (e.g., fiberoptic deployments, robotic drills, subglacial samplers) could transform scientific understanding. He noted that the United States lacks an operational capability to access the ice-bed interface and that hot water or rapid drills may be necessary for IPY5 subglacial access but might not be ready by then.

Power supply was characterized by Siegfried as the linchpin for autonomous stations and cleaner, scalable systems for field camps, and he observed that diesel generators still dominate large camps and recounted hauling heavy lead-acid batteries across Antarctica to power instruments. Siegfried suggested that lower-mass, hybrid renewable systems and smarter power management can extend endurance and reduce logistical burden. IPY5, he noted, could be an opportunity for "an energy transition" for polar science.

PERMAFROST, FIRE, AND CARBON

Brendan Rogers, Woodwell Climate Research Center, outlined three problem areas where IPY5 could accelerate progress: abrupt permafrost thaw (e.g., thermokarst), wildfire, and the overall Arctic carbon balance.

Permafrost

Rogers described permafrost change as both local (e.g., damaging foundations, roads, and ecosystems) and global (e.g., carbon feedbacks). Abrupt thaw features, such as thermokarst lakes and ground collapses, are widespread yet unevenly mapped; circumpolar mapping could be accelerated by high-resolution commercial imagery coupled with artificial intelligence (AI), but access, cost, storage, and computing constraints limit scale-up. Repeat lidar, interferometric SAR, and similar tools can detect subsidence and deformation, and expanded ground-truth data—validated feature catalogs and in situ carbon flux and transport measurements in abrupt-thaw terrains—can anchor interpretation. Rogers noted that Earth system models underrepresent abrupt thaw processes and scale mismatches remain a central hurdle; targeted model development and parameterization, informed by new observations, can improve projections.

Arctic Wildfires

Rogers observed that Arctic–boreal fires are intensifying with climate change, with implications for health, Indigenous cultural values, infrastructure, permafrost stability, and the carbon cycle. Knowledge is comparatively strong for Alaska and Canada but thinner for Siberia. He pointed to opportunities to link field measurements (e.g., ground-penetrating radar, eddy-covariance towers) with repeat satellite mapping to quantify fire impacts on permafrost and improve emissions inventories, transport modeling, and community-relevant forecasting tools. Fire management, he suggested, can blend emerging technologies (e.g., enhanced fire-detection satellites, unmanned suppression platforms) with revitalized cultural burning practices that have long reduced catastrophic risk.

Carbon Budget

Turning to the carbon budget, Rogers noted that present-day Arctic net carbon balance estimates show limited agreement, and potential future releases remain large. The eddy-covariance tower network directly measures carbon dioxide (CO_2) and methane (CH_4) exchanges and is expanding (e.g., Permafrost Pathways, the Next-Generation Ecosystem Experiments, FLUXNET); sustaining and extending it—especially in Siberia and disturbed ecosystems—can enhance constraints. He suggested modeling priorities (e.g., improved snow physics, terrestrial–aquatic linkages, dynamic vegetation) and emerging satellite approaches, including column CO_2/CH_4 measurements and lidar concepts that can operate through the polar night. The strongest gains, he suggested, may come from coupling "bottom-up" data (e.g., flux towers, field campaigns, high-resolution mapping) with "top-down" atmospheric inversions to reduce uncertainty.

Rogers closed by underscoring how this work unfolds on Indigenous lands. Coproduction of knowledge, respect for Indigenous data sovereignty, and community priorities can be central to an effective IPY5 portfolio.

AUTONOMOUS OCEAN OBSERVING SYSTEMS

Progress Since IPY4

Esmee van Wijk, Commonwealth Scientific and Industrial Research Organization in Australia, reviewed broadscale ocean observations from Southern Ocean Argo since IPY4 (see Figure 5-1). Ice-sensing algorithms, data storage, and faster, two-way satellite communications have increased float survivability in seasonal ice, enabling year-round subsurface measurements. Despite serious progress in filling the huge data gaps in the Southern Ocean, significant gaps remain. The deep ocean below 2,000 m is sparsely observed. Deep Argo can address transitioning from regional arrays toward global implementation. Biogeochemical Argo (BGC Argo) has expanded and added new sensors that can help close the carbon budget in the Southern Ocean. Van Wijk underscored that parts of the seasonal ice zones (e.g., Ross, Weddell) and continental shelf around Antarctica, are still sparsely observed. Only about 100 floats have been deployed on the latter. She suggested the need to scale this to a circumpolar array and noted ongoing discussions about whether Antarctic shelf floats should become part of core Argo.

Under-Ice Operations and Multiplatform Systems

Van Wijk described significant advances in observing gliders and other autonomous underwater vehicles that now operate for months to years over hundreds to thousands of kilometers, with under ice-missions reaching beneath shelves. Uncrewed surface vehicles (e.g., wave gliders, sail drones), long-range drones capable of deploying instruments in remote regions, noting recent Arctic demonstrations of basin-scale autonomy and acoustic networking (Lee et al., 2022). International collaboration can scale up these approaches, drives advances in understanding, and maximizes the use of resources for collective benefit.

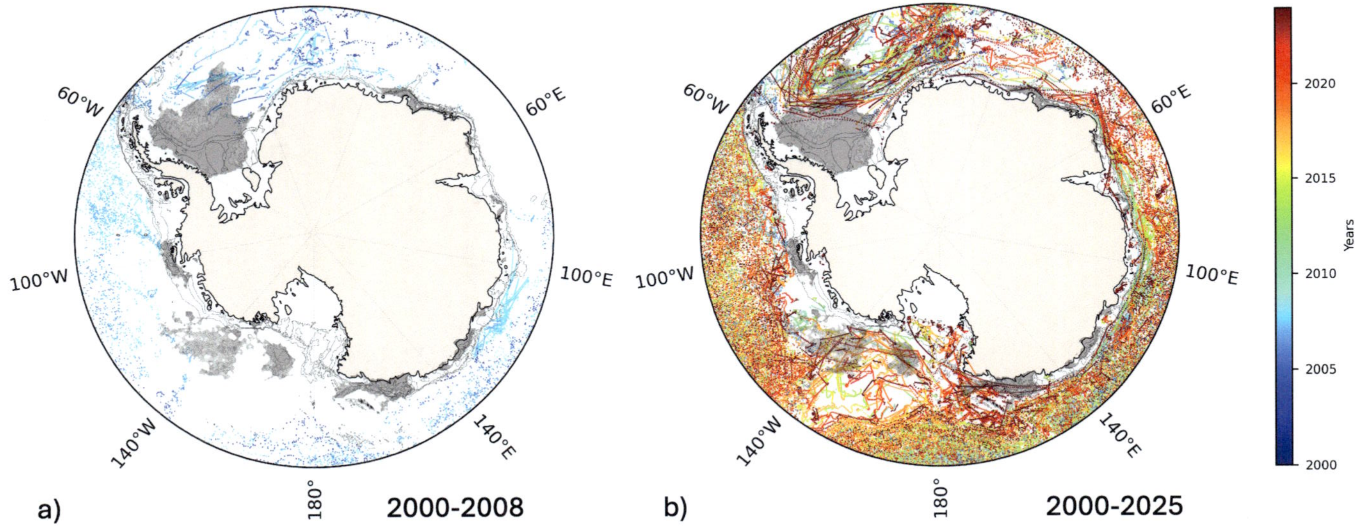

FIGURE 5-1 Broadscale observations of the Southern Ocean compared between 2008 and 2025.
NOTE: Broadscale ocean observations from Argo float profiles in the Southern Ocean (south of 60°S), color-coded by time of observation. Panel a): Argo float profiles measured 2000–2008; panel b): Argo float profiles (2000–2025) highlighting increased coverage since IPY4.
SOURCE: Presentation by Esmee van Wijk on May 21, 2025; figure prepared by N. Kolodziejczyk.

Examples of Potential Next Steps

Van Wijk pointed to gains made through developments in cost-effective, lower-power, longer-lived platforms and sensors; expanded long-range geolocation, networked for communication; and data telemetry under ice. She suggested that opportunistic sampling and deployments from tourist, fishing, and merchant vessels (including bathymetric data collection) would benefit from expansion. She underscored the need to address Southern Ocean cloud biases by measuring aerosols, radiation, and precipitation from ships and combining these with satellite observations, to develop better model parameterizations. Ships and BGC Argo floats provide gold-standard observations needed to understand the carbon cycle in the Southern Ocean. Process studies are still required to understand physical, biogeochemical and biological coupling, and carbon uptake variability. Van Wijk said that IPY5 could provide a springboard to scale up to an integrated, international, multiplatform, and multidisciplinary observing system that could better constrain the Southern Ocean's role in climate and sea-level rise.

MODELING AND MACHINE LEARNING

Charles Koven, Lawrence Berkeley National Laboratory, focused most of his presentation on Department of Energy projects he has been involved with. He began with a reminder that the Keeling Curve, launched during the 1957–1958 International Geophysical Year, sets a high bar for lasting IPY legacies. Looking toward IPY5, he observed that CO_2 trajectories are diverging (e.g., Paris Agreement–aligned reductions versus accelerating human-caused CO_2 emissions growth), but in either case, the Arctic is poised for dramatic change through the 2030s. He spent the remainder of his time on modeling carbon cycle feedbacks.

Process Representation and Ensembles

Koven focused on carbon cycle feedbacks under divergent CO_2 trajectories and observed that many Earth system models, such as those from the Intergovernmental Panel on Climate Change, omit critical Arctic processes, including fire, vegetation dynamics, permafrost carbon processes, and the nitrogen cycle. In particular, permafrost carbon—missing from almost all models—means feedback between fire, vegetation change, and permafrost is not captured, underscoring that this omission is "the core modeling challenge," he said. Koven then described fine-scale models (e.g., polygonal tundra hydrology)—as ice wedges melt, polygons subside, drainage shifts, and ponding patterns change, fundamentally altering thaw rates and carbon emissions (see Figure 5-2). He suggested that both offline fine-scale models and coupled models, run in large, perturbed parameter ensembles, can expose system behavior and quantify uncertainty.

Data and Machine Learning

Koven argued that increasing land model complexity benefits from better constraints and that machine learning can replace empirical parameterizations (e.g., phenology) and serve as fast emulators for large ensembles. He pointed to emerging benchmarking—scorecards comparing models to observations across a wide array of variables, such as biomass, permafrost area, ocean fluxes—and suggested that IPY5 could support systematic benchmarking to track improvement. Retaining research capacity and institutional support was noted as important for continuity.

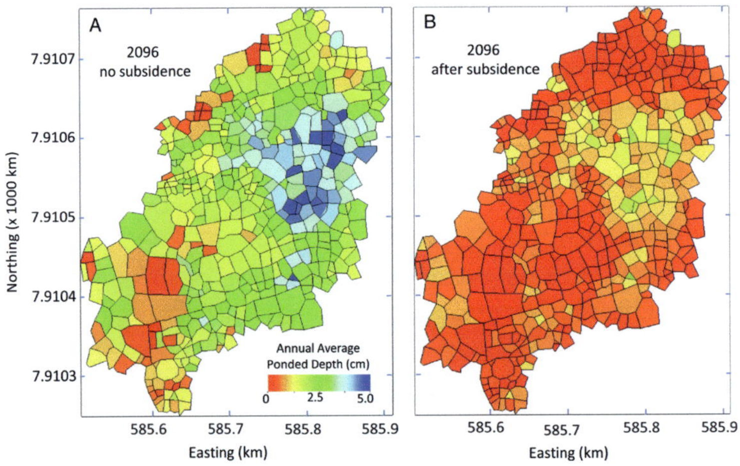

FIGURE 5-2 Incorporating Fine-Scale Permafrost Thaw Processes into Land Models.
NOTE: Modeled changes in annual average ponded depth illustrate the incorporation of a key mechanistic fine-scale process into land models.
SOURCE: Presentation by Charles Koven on May 21, 2025; Painter et al. (2023). CC BY 4.0.

POLAR EARTH OBSERVING NETWORK (POLENET) AND LESSONS FROM IPY4

Terry Wilson, Ohio State University, focused on specific lessons and successes with her experience on a NSF–funded IPY4 project, POLENET. Wilson described how years of pre-2007 coordination through the Scientific Committee on Antarctic Research (SCAR) and U.S. facility partners (e.g., NSF, NASA) enabled rapid deployment of autonomous GPS and seismic stations once IPY was launched. Before, she explained, only a handful of such systems existed. During IPY4, many nations funded deployments, and "we achieved a step change in polar observations across Greenland and Antarctica." She emphasized practical design elements that made those networks durable and scalable in remote sites—sensors, power, communications, survivability (i.e., hardiness), transportability, ease of assembly, and cost control—and illustrated them with concrete examples (e.g., long-lived field power built from banks of car batteries, iridium telemetry for near-real-time GPS data), underscoring that the same technology is still being used with only incremental improvements because the core concepts have proven durable.

Wilson also highlighted legacies that continue to be useful beyond IPY4: open data pipelines that enable broad reuse (including by communities outside the original discipline), training schools structured to seed international peer networks for early-career participants, and partnerships that lowered costs, such as U.S. Antarctic Program negotiations that reduced iridium charges on the condition that data flowed to public archives. Looking ahead, she observed that shared logistics and pooled, multinational mechanisms can unlock larger, longer-lived efforts—including public–private collaborations (e.g., privately financed airborne radar later shared with researchers)—and suggested that "plug-and-play" field backbones for power and communications can support multisensor, multiproject sites and reduce duplication.

Wilson suggested the following potential actions ahead of IPY5:

- Build an international community, and articulate priority science objectives; initiate technology development, and coordinate around performance metrics and operational best practices.
- Advance autonomous sensor systems, including developing (which may be necessary) power and communication systems tailored to polar conditions.
- Consider multisensor platforms (requirements, feasibility), and coordinate data acquisition modes so that products serve multidisciplinary science.
- Run workshops and training schools to foster international early-career researcher (ECR) networks positioned to lead IPY-scale projects.

- Pursue mechanisms for shared funding of science, logistics, and communications across nations.

Wilson's operational emphases included the following:

- Open data usable across disciplines to maximize reuse and synthesis.
- Shared communication systems and cost sharing to stretch resources and support real-time operations where appropriate.

Together, Wilson's reflections point to the following options for IPY5: investing early in shared backbones (e.g., power, communications, open data); scaling what already works; and pairing technology deployments with deliberate, international ECR training—so that by the time IPY5 launches, both the hardware and human networks are ready to act at scale.

DISCUSSION

Crosscutting Observing Strategies

Schmidt, the panel's moderator, opened with the following question: "are there particular observing strategies or technologies that could serve multiple disciplines at once—for example, ice sheet science, permafrost, and ecosystems—to make IPY5 investments more impactful?"

Siegfried emphasized the value of "4-D data"—3-D measurements repeated through time—arguing that a swath-capable spaceborne lidar with weekly or biweekly revisits could capture episodic change across ice sheets, permafrost landscapes, and terrestrial ecosystems. Rogers added that commercial high-resolution imagery paired with AI already delivers a step change for mapping abrupt thaw and thermokarst while simultaneously detecting vegetation shifts, fire scars, and hydrologic change—benefits that expand as access and cost barriers are addressed.

From the ocean perspective, van Wijk pointed to under-ice acoustic geolocation, networked for communications and data telemetry, as shared infrastructure to enable a broad range of autonomous platforms—an approach she said that could multiply returns when deployed at scale. Speaking to modeling, Koven noted that interoperable, common observational products underwrite machine-learning benchmarks and allow whole-system integration across ocean, ice, permafrost, and ecology. Wilson underscored the value of ground-based "plug-and-play" stations—shared power and communications backbones onto which diverse sensors (GPS, seismic, magnetics, weather) can be added—thereby reducing duplication and leveraging field logistics. An online participant observed the lack of the atmosphere and key atmospheric processes; the participant noted that a major challenge for global modeling is, for example, Southern Ocean clouds, and understanding how to better integrate atmospheric research needs and linkages into discussions about IPY.

Powering Polar Research

In response to a prompt about power for IPY5, Siegfried argued for moving away from lead-acid batteries toward renewable, autonomous, scalable station power; he suggested that early investment in hybrid/clean systems could displace diesel at field camps and give stations multiyear endurance. Wilson noted that batteries from 2007 are still operating—evidence that incremental improvements can pay off—but pointed to gaps where higher-bandwidth communications and smarter power management could markedly improve performance.

International Coordination

Schmidt pivoted to international coordination and van Wijk pointed to Argo as a working model: common standards, high-quality data, open data sharing, and distributed national contributions that could be extended to include polar shelf floats, or used as a model for gliders, and moorings. From the modeling side, Koven highlighted the value of coordinated intercomparisons and shared benchmarks, like the Coupled Model Intercomparison Project, to help national efforts converge scientifically. Rogers emphasized partnering with Indigenous communities from the outset—framing "international" to include both peoples and nations—so that shared investments align with local priorities and knowledge systems.

Together, the exchanges by speakers during the discussion suggested a dual track for IPY5: modernize field power systems while aligning international standards, open data practices, and inclusive partnerships to sustain multi-country efforts over time.

INTERACTIVE ACTIVITIES: APPROACHES AND CAPABILITIES TO SUPPORT IPY5

Session 3 concluded with an interactive activity in the style of the World Café method (in-person attendees) and breakout discussions (virtual attendees).[1] At-

[1] More information about the World Café method is available at https://www.involve.org.uk/resource/world-cafe.

tributions in this section indicate the reporter for each group; individual points reflect what they said on behalf of their groups.

Question 1: New or Emerging Technologies or Capabilities

Kirsty Tinto, Lamont-Doherty Earth Observatory at Columbia University, reported interest in winter-hardy autonomous sensors, distributed instrument kits, and fiber-optic distributed acoustic sensing, paired with AI to manage and triage large data streams; the discussion also pointed to improved atmospheric profiling in stratified layers, citizen science, and community partnerships as testbeds for technology. Dave Sutherland, University of Oregon, added that model–observation iteration is important, AI can be powerful when its assumptions are transparent, and a "digital twin" of the polar oceans—echoing EU efforts—could be explored alongside low-cost greenhouse gas sensors.

Andrew Lloyd framed "how fast, and by how much, will the ice sheets melt?" as a central, crosscutting question with clear societal stakes. He emphasized two-way coupling between models and observations—modelers can flag the measurements most likely to reduce uncertainty, and observers can target those in the field. Lloyd also pointed to undersampled regions (notably East Antarctica), ice–ocean interactions beneath shelves, landscape change (e.g., permafrost thaw, fire), and human activity (e.g., tourism) as areas where additional work can inform local decision tools. He noted field safety and inclusivity as integral to future efforts.

Question 2: Novel Tools or Approaches to Consider, Including from Disciplines/Applications not Typically Used in Polar Research

Lindsay Goodwin, New Jersey Institute of Technology, relayed caution about relying on unproven tools; some participants favored scaling proven technologies while launching long-term studies, strengthening disease surveillance (e.g., avian influenza), and engaging Alaska regional hubs and local expertise. Scott Goetz highlighted swarms of drones as novel tools, greater use of commercial satellite imagery, and selective access to declassified datasets.

Mary-Louise Timmermans, Yale University, highlighted the value of including Indigenous peoples early in planning and suggested that a large U.S. effort focused on constraining sea-level rise could be high impact. She also mentioned permafrost questions (e.g., carbon and methane emissions), core drilling opportunities, leveraging International Arctic Research Policy Committee planning, and using tourism and shipping as observing platforms. Timmermans noted growing interest in AI for remote sensing and urged careful consideration of the ethical and social dimensions of geoengineering concepts (e.g., engineered ice growth).

Ginny Catania, the University of Texas at Austin, relayed questions that link the Arctic and Antarctic: how past polar change unfolded; where processes overlap or diverge (e.g., the "Greenlandification" shorthand for Antarctic change); and the teleconnections that produce synchronous or asynchronous responses. She conveyed the view from Martin Jeffries, chair of the Polar Research Board, that IPY5 could deliver improved climate projections that inform adaptation and mitigation, coupled with tangible products and services that go beyond traditional outputs, like a journal article.

Question 3: The Role of International Collaboration in Fostering the Development and Use of New Capabilities and Approaches and Experts/Groups Involved in Fostering Relationships

Oscar Schofield, Rutgers University, described two complementary coordination modes: grassroots working groups through SCAR/International Arctic Science Committee/Arctic Science Summit Week and top-down steering teams modeled on efforts like the NASA Arctic Boreal Vulnerability Experiment, suggesting a shared calendar so that early-career and community members can plan access to common assets. Melissa Shapiro, Woodwell Climate Research Center, emphasized programmatic coordination to close gaps and address inequities (e.g., English-language dominance), along with early, sustained partnerships with Indigenous communities and cross-sector participation (e.g., policy, communications, community liaisons).

Jenny Baeseman, planning committee member and polar consultant, reported broad topic lists across groups (e.g., ice sheets, sea level, ecosystems, submarine melt, magnetosphere, forecasting, teleconnections), frequently emphasizing closing the carbon budget, and mentioned including the "third pole"—the Himalayas—in comparative work. Several groups encouraged focusing on activities possible only at IPY scale rather than efforts typical of smaller programs.

Lilian Alessa, planning committee member from the University of Idaho, described environmental services and operational needs: leveraging and sharing existing resources; lower technology costs to broaden participation; engaging communities beyond the usual networks; and exploring dual use opportunities (e.g., sea-ice monitoring that supports both science and security). She suggested that the global polar science community is larger than often assumed and could benefit from clear taxonomies for collaboration that weigh risk and benefit.

Jamin Greenbaum, planning committee member from the Scripps Institution of Oceanography, underscored multiplatform deployments that combine high-frequency radar, Argo floats, seismics, and other assets, supported by standardized field kits and coordinated use of surface, airborne, and satellite platforms. He cited programs such as the expedition Multidisciplinary Drifting Observatory for the Study of Arctic Climate, European Space Agency mission Swarm, and POLENET as precedents and observed that creative, integrated data systems can help reduce disciplinary silos.

Question 4: Technologies or Approaches where Researchers Can Inform Development to Support IPY5

Byron Adams, Brigham Young University, underscored co-located permafrost–ecology monitoring, integration of paleo and modern observations for synchronized global synthesis, and Arctic–Antarctic synthesis, including human biology and inviting partners beyond the usual circle (e.g., fisheries managers, virtual reality [VR] specialists). Adams also emphasized coupled atmosphere–ocean–ice studies; co-located biogeochemistry and ecosystem surveys; moorings that combine weather, biological, and seismic sensing; and permafrost–ecology sampling. He also noted human biology and the opportunity to tap underused networks (e.g., SCAR) and bring in partners such as fisheries managers and VR specialists. Greenbaum echoed the value of multiplatform deployments (e.g., balloons, moorings, seismic, Argo) and pointed to interoperable data systems as a practical way to reduce silos across projects.

Ha Hoa Hamano, NASA, summarized environmental themes, such as wildfire monitoring; long-term ocean observing around ice sheets; subglacial access infrastructure; and Arctic pollutants, alongside enabling structures such as enhanced instrument connectivity, support for Indigenous data sovereignty, unified mapping tools, coordinated satellite systems, and remote base camps for multidisciplinary work. She underscored the range of expertise that can contribute, from physicists and geospace researchers to policy experts, Indigenous partners, and citizen scientists.

In closing, Craig Lee noted the challenge ahead: shaping these ideas into a structure that can argue for ambitious programs, with clear pillars and unifying concepts for IPY5.

6

Recap of Day One

Day 2 opened with a recap of Day 1 to draw out crosscutting "umbrella" themes and prompt participants to contribute further ideas via the online Q&A platform. Planning committee chair Craig Lee framed the goal of the workshop as looking beyond specific projects to the larger structures that could anchor an International Polar Year (IPY)–scale effort.

LINKING SCIENCE TO SOCIETAL OUTCOMES

Lynn Talley, University of California—San Diego, provided a structured, personal summary of Day 1. Talley emphasized that IPY5 can explicitly connect polar science to services and products for society, such as, forecasts, decision tools, and sustained data streams, alongside foundational publications. She suggested weaving observations, modeling, and rapidly maturing artificial intelligence methods into one programmatic fabric, so that field campaigns, remote sensing, and model development iterate in near real time and yield decision-quality outputs. Talley's summary captured research and monitoring areas addressed throughout Day 1, including sea-level and ice dynamics, space weather, permafrost carbon, and ecosystem change, where speakers repeatedly tied science advances to community needs and operational planning.

However, Talley flagged ecosystem shifts as one of the core science topics less represented during Day 1 and urged that IPY5 efforts fully incorporate ecosystems into Earth-system models so that forecasts and products connect physical change with ecological responses. Participants in the online chat also pointed to the importance of circumpolar ecosystems, and large-scale impacts of both regions on global climate. Additionally, Talley also noted underrepresentation of atmospheric processes and atmospheric science perspectives during discussions at the workshop, a point echoed by several participants in the online chat. The following sections integrate Talley's recap with comments occurring from the audience and in the online chat.

Big Questions and "Umbrella" Topics

During the recap, Tally noted that several "big-picture" research areas emerged as focal candidates for IPY5: (1) sea-level rise and grounding-zone processes; (2) sea-ice loss and coupled atmosphere–ocean responses; (3) permafrost thaw and the Arctic carbon budget; (4) fire in northern high-latitude ecosystems; and (5) space weather impacts on communications, power, and aviation. Participants in the online chat also pointed to ecosystem shifts and human dimensions—the translation of global results to local decisions, tourism, and other human activities in polar regions—as integral threads within these areas. Lee proposed that a small set of well-scoped umbrella topics could help align proposals, coordination mechanisms, and shared infrastructure during the compressed IPY5 window.

Approaches and Capabilities

Looking ahead to "how," Talley highlighted three programmatic pillars: (1) sustaining and extending observing systems so that IPY5 acts as an intensive phase embedded in longer records; (2) accelerating model improvement—spanning process representation, ensembles, and systematic benchmarking—codeveloped with targeted observations; and (3) strengthening data management so that observations and model products are openly available and interoperable. She observed that many technologies and networks exist (e.g., autonomous platforms, acoustically enabled arrays, shelf/shelf-break and under-ice assets); the opportunity is to scale, connect, and standardize them for multidisciplinary use.

Participation, Inclusion, and Partnerships

Multiple comments from Talley and online participants during the recap underscored early, substantive roles for early-career researchers and Indigenous partners in shaping IPY5. Suggestions included tracking who participated in planning (e.g., career stage, discipline) to clarify which perspectives were included; embedding Indigenous data sovereignty practices; and bringing in communicators, artists, and policy liaisons as part of the core team to broaden reach and coproduce science. Several participants also pointed to funding and coordination models that span agencies and sectors, noting that pooled or harmonized arrangements (e.g., international, interagency, public–private) can amplify observing and modeling investments beyond standard grant cycles.

Timing, Scale, and Realism

Talley said that several participants observed that 2032–2033 is fast approaching in program terms; IPY5 can build on existing assets and planning efforts (e.g., Interagency Arctic Research Policy Committee, Southern Ocean initiatives), expand what already works, and pilot pragmatic enhancements rather than relying on unproven technologies. Several unique near-term opportunities were also flagged (e.g., a strong solar maximum and total solar eclipse over Alaska in 2033) that could anchor compelling geospace and public-engagement activities. Links to planetary missions (e.g., Europa Clipper) were added as cross-community points of synergy.

Framing for Impact

Talley's closing point was practical: IPY5 can provide tools—forecasts, warnings, shared situational awareness, and decision supports—for communities and sectors while advancing core science. In this framing, success could rest on selecting clear scientific foci, aligning observations and models to those foci, and ensuring the resulting systems persist beyond the 2-year IPY5 intensive phase through coordinated international follow-on activities.

Lee transitioned the meeting to the Day 2 focus on human capacity and collaboration, noting that the people, skills, and partnerships that sustain these systems are as important as the technologies themselves.

7

Human Capacity and Collaboration

Day 2 emphasized people and partnerships, signaling a shift from what to study toward how the work is done and sustained. The day started with Session 4, Human Capacity and Collaboration, which underscored that advancements in understanding of the polar regions are underpinned by expansive teams that extend well beyond the principal investigators (PIs) or core team for any given project. This can include community members, tourists, and the public who contribute to data collection in their communities or aboard cruise ships; coordinators who manage complex logistics for deploying personnel and equipment; pilots, captains, and drivers who transport researchers to remote sites; and engineers and technicians who build and maintain specialized equipment and models, among others. The session also looked at some specific examples of how research is conducted and the practicalities and logistics required.

Lil Alessa, planning committee member from the University of Idaho, moderated the first panel of the session, Strengthening Human Capacity. The panelists were prompted with the following discussion questions:

- What do you see as the primary strengths and challenges of supporting polar research today?
- Do you envision supporting International Polar Year (IPY) 5? If yes, what might that look like? Do you expect strengths and challenges to change moving forward to IPY5?

SUSTAINING HUMAN CAPACITY: WORKFORCE, LOGISTICS, AND TRAINING PATHWAYS

Ed Brook, Oregon State University, opened the first panel by describing the Center for Oldest Ice Exploration, the National Science Foundation (NSF) Science and Technology Center he directs, which aims to recover very old Antarctic ice and grow a broad community around education, knowledge transfer, and participation. He emphasized that access to the ice sheet interior depends on complex logistics (e.g., aircraft, traverses, cold-chain curation of cores) and that it has become more difficult; relying on aging systems, he suggested, may constrain both science and safety. For human capacity, he emphasized that such distributed programs can serve as "microcosms" for IPY5 partnerships—linking research, training, and logistics across many institutions—and stage the next set of international efforts as the IPY5 window approaches.

Turning to people, Brook underscored the cumulative, specialized expertise required for field success—scientists, engineers, pilots, drillers, cargo handlers, field guides, camp managers, and technicians. He observed that many on field teams hold time-limited positions, raising questions about long-term retention and career security. He suggested explicitly describing human capacities needed; creating pathways that keep people in polar work (internships, post-baccalaureate roles); cross-training with international partners; embedding training and workforce development in logistics contracts; and fostering partnerships among science teams, logistics providers, and nongovernmental organizations.

PROTECTING PEOPLE AND PARTNERSHIPS

Twila Moon, National Snow and Ice Data Center, framed near-term risks to agencies, budgets, and personnel as a human-capacity issue, cautioning that rebuilding strained institutions can be challenging. She suggested concretely engaging with policy processes and philanthropy (e.g., using established "stand up for science" toolkits) and "connect[ing] the dots" for decision makers by linking polar change to local impacts and services. Relationship-building, especially with communities and organizations on the ground, was presented as a core practice that can maintain communications through periods of geopolitical or institutional stress.

Moon highlighted practices that can make science more usable by aligning with the principles of (1) findability, accessibility, interoperability, and reuse of digital assets; (2) collective benefit, authority to control, responsibility, and ethics) for data; and (3) transparency, responsibility, user focus, sustainability, and technology for data repositories. She also highlighted creating products that practitioners can act on, such as policy briefs like Arctic Answers[1]—two-page briefs that answer questions about Arctic environmental change—and other planning tools; and bringing in communicators and artists to help connect evidence to decisions. She suggested moving from consultation to codevelopment (e.g., practitioner boards that guide tool development to ensure products and results are usable) and noted the value of low-cost, community-operated observing tools where appropriate. A recurring theme was prioritizing people: funding whole teams (not only PIs), cultivating a culture where researchers can share constraints and ask for help across groups and institutions, and reducing reliance on unpaid labor, in part, through more usable administrative systems.

CENTERING INDIGENOUS CAPACITY: RELATIONAL SCIENCE, DATA SOVEREIGNTY, AND LOCAL MONITORING

Hannah-Marie Garcia-Ladd, Indigenous Sentinels Network (ISN), opened by noting that "data is not a foreign concept" in Indigenous contexts: many communities have practiced systematic observation for generations to support safety and stewardship. She proposed reframing "capacity" to include the people who live with and operate in the polar landscape year round and those who run logistics and support systems. From the vantage of the ISN, which began on St. Paul Island and now supports dozens of Alaska communities, she described a model that pairs community-owned data systems (e.g., tiered access, trusted channels, explicit attribution) with programmatic support (e.g., grant writing, science communication/storytelling, training in observing methods, drones, analytics).

Examples included partnerships that connect local mariners to operational services (e.g., freezing-spray observations with the Alaska Ocean Observing System to inform weather forecasts) and efforts such as Skipper Science that link fishermen to real-time environmental monitoring. She suggested that embedding Indigenous data sovereignty and technology transfer into IPY5 planning and implementation can strengthen both science and stewardship. In closing, Garcia-Ladd summarized three points: (1) local monitors are relational scientists, not just participants; (2) Indigenous data sovereignty and tech transfer can be built into programs from the outset; and (3) durable capacity building can include workforce training, youth engagement, and support for community technicians and guardians.

[1] More information about Arctic Answers is available at https://www.tandfonline.com/journals/uaar20/collections/Arctic-Answers

DISCUSSION

Engaging the Next Generation

In response to a question about reaching today's 15-year-olds—who could be first-year graduate students during IPY5—Brook pointed to teacher engagement and "polar literacy" resources as near-term avenues; Moon noted that mindful attention to early-career on-ramps is warranted, including outreach in local high schools; and Garcia-Ladd added that tribal governments are building labs and research centers to introduce local youth to hands-on science.

Barriers to Networks

A participant raised the challenge of breaking into professional networks without established mentors or institutional pedigree. Moon observed that "open" mechanisms (e.g., a public call to join a committee) are not automatically "welcoming" and encouraged explicit invitations, clear involvement pathways, and follow-through. Brook described how Science and Technology Centers can bring early-career researchers (ECRs) into proposal development and field planning, while acknowledging that this is added work for both them and senior researchers and may require explicit time and budget. Garcia-Ladd emphasized that community partners face heavy request loads; patience and relationship-building can align asks with local priorities.

Actionable Science and Agencies

Alessa asked how "actionable science" that serves agencies and communities can help stabilize capacity. Moon suggested collaborations designed from the start with end users (e.g., practitioner boards in coastal sea-level work), localizing information (from national projections to place-based services), and planning for operationalization beyond the discovery phase. Some participants also mentioned early-career self-organization (e.g., Association of Polar Early Career Scientists [APECS] and Polar Early Career World Summit networks) as a way to identify priorities that persist beyond individual postdoctoral terms.

STRENGTHENING RESEARCH CAPACITY—BUILDING ON LESSONS LEARNED

Framed within "the emphasis that Day 2 placed on people, partnerships, and sustained collaboration, the next panel explored what it takes to mount ambitious polar efforts, campaign design, leadership development, data systems, and community-based observing, and how lessons since IPY4 can inform IPY5 planning. The session emphasized enabling leadership (including ECR roles), durable logistics and data infrastructure, and equitable engagement with local and Indigenous partners. Mike Hartinger, planning committee member from the Space Science Institute, moderated the discussion. Panelists were provided with the following discussion questions:

- What are the primary strengths and challenges to the current approach to conducting research and supporting ECR development? What lessons can be learned from prior campaigns?
- These activities require many skilled individuals beyond the researchers themselves. Are there key areas where capacity development will be required to support IPY5-level activity? How might we do this?
- What opportunities does the ramp-up period to IPY5 provide to strengthen this capacity?
- Are there mechanisms to help sustain the needed capacity over time?

LEADERSHIP, INFRASTRUCTURE, AND CROSS-DISCIPLINARY PRACTICE FROM MULTIDISCIPLINARY DRIFTING OBSERVATORY FOR THE STUDY OF ARCTIC CLIMATE (MOSAiC)

Matthew Shupe, University of Colorado—Boulder, kicked off the panel by describing the year-long MOSAiC expedition—an icebreaker frozen into Arctic sea ice that required ~20 nations, hundreds of personnel, multiple ships, and hundreds of millions of dollars—as a testbed for building and sustaining capacity at scale. He said that successful "big science" hinges on cultivating and enabling leadership (people empowered and resourced to convene communities, manage details, and sustain momentum), designing inclusive roles for ECRs in the science, fieldwork, synthesis, and even leadership, and resourcing the coordination work that often goes unfunded. He emphasized large-scale, shared infrastructure—from ships to observing nodes—as a nucleus that draws in diverse science and lowers marginal costs for participation.

Shupe emphasized that capacity grows when teams plan across disciplines, such as agreeing up front on shared questions, products, and synthesis plans, while maintaining a strong safety culture and committing to open data. He said that securing the "big pieces" early (e.g., ship time, aircraft hours, access to key stations) provides a stable foundation for partners to align and build ambitious work. He urged stronger interoperability so that observing networks and modeling groups use common formats, application programming interfaces (APIs), and workflows, letting observations flow directly into models and comparative analyses. He also called for near-real-time, open data sharing during field campaigns—posting provisional, well-documented data and quick-look products—so that analysts not in the field can troubleshoot, cross-validate, and contribute insights that accelerate discovery. Finally, he pointed to power and autonomy as immediate levers—hybrid/renewable power, smarter power management, and more autonomous platforms—to lengthen deployments and reduce logistics loads.

SUB-ICE DRILLING AS CAPACITY-BUILDING SCIENCE

Ryan Venturelli, Colorado School of Mines, outlined scientific targets that call for scalable, modular, clean-access drilling and long-term observing beneath ice shelves and grounded ice, linking tidal-to-millennial variability at grounding zones to projections of retreat and sea-level rise. She described the Subglacial Antarctic Lakes Scientific Access project at Mercer Subglacial Lake as a model: a transdisciplinary team, including expertise from microbiology, geochemistry, geology, geophysics, glaciology, and drilling engineers, that paired high-quality science with deliberate mentorship and leadership growth for students and postdocs, plus public communication (e.g., film). In her view, sub-ice access campaigns can double as leadership pipelines, technology incubators, and community builders.

Venturelli suggested that IPY5 could emphasize modular, forward-compatible systems, such as drills, observatories, and sensors that can be adapted, redeployed, and extended, with clear international leadership and accountability frameworks. She also noted that whenever holes are drilled, leaving the instruments behind (e.g., distributed acoustic sensing, moorings) can generate sustained time series that outlast a campaign and support models that track small-scale processes up to large-scale sea-level outcomes.

SCIENCE-READY, FINDABLE, AND REUSABLE DATA

Katherine Cariglia, Massachusetts Institute of Technology Haystack Observatory, presented on Madrigal, an open-source, distributed data system widely used in geospace and upper-atmosphere research, as a template for science-ready data. It uses standardized formats and units, self-describing metadata (e.g., parameters, uncertainties, quality flags), citable records, and multiple access modes (e.g., web user interface, Python/MATrix LABoratory/Interactive Data Language, APIs). Madrigal hosts 83 TB of data from 200+ instruments and ~300 TB of backups, supports real-time and archival data streams, and allows producers to upload directly or host their own node within the community catalog, keeping data local but discoverable and retrievable through a single portal, which reduces effort for both contributors and users.

From a capacity perspective, Cariglia emphasized usability and sustainability: making it easy to contribute well-documented data; investing in tools that help users find, filter, and combine datasets; and supporting the people who operate archives and assist contributors. She pointed to summary plots, dataset verification by originators, and consistent metadata as practical steps that increase reuse across disciplines (e.g., pairing ionospheric disturbances with atmospheric measurements or radio propagation with weather and space-weather conditions).

COMMUNITY-BASED OBSERVING NETWORKS (CBONS) AND TRUST

Andrew Kliskey, University of Idaho, described CBONs as a scalable, durable way to strengthen IPY5 capacity—linking local observers, technology, and data science. He traced examples from the Bering Sea Sub-Network and its successor, Community-Based Observing Network for Adaptation and Security, which documented climate and biological change through observations by Indigenous hunters, fishers, and other place-based experts. He situated CBONs within a spectrum—observer forums (e.g., Local Environmental Observer Network), citizen science, community-based monitoring, and CBON systems—each with different strengths on data accessibility versus interoperability.

Kliskey underscored the role of trust, continuity, and bidirectional data flow: tools that return synthesized information quickly to observers; data pipelines designed for austere environments; and governance that respects Indigenous data sovereignty. He also noted that methods and relationships established in the Arctic are now informing community-based observing in other regions (e.g., wildfire and drought contexts), suggesting that CBONs can link local insight to regional and national networks if supported over decadal timescales.

DISCUSSION

Early-Career Leadership and Roles Beyond Academia

Panelists and many participants emphasized placing ECRs in visible, consequential roles, including coordinating field teams, leading synthesis papers, and shaping data products. The discussion also highlighted the broader workforce (e.g., engineers, software developers, technicians, logistics specialists) as core to polar capacity; teams can plan budgets and review narratives that recognize these roles, rather than assuming one-PI/one-student models.

Data Accessibility and Open Sharing

Participants pointed to the value of early, open sharing of data and research outcomes during campaigns, such as what was used in MOSAiC, to invite wider analysis, while acknowledging concerns about misuse or credit. Several participants exhibited strong interest in science-ready metadata, consistent formats, and interoperable repositories paired with user support and tools that make data understandable to broader communities as well as scientists.

Codes of Conduct, Safety, and Culture

Multiple participants mentioned field safety and conduct, not only written policies but also clear expectations, reporting paths, consequences, and attention to cross-cultural contexts on multinational teams. Existing community codes (e.g., professional society guidance, shipboard conduct frameworks) were cited as starting points; some participants suggested that an IPY5 framework could harmonize standards and clarify enforcement across funders and institutions.

International Coordination and Shared Assets

The discussion returned to shared infrastructure (e.g., ships, camps), smarter power systems for long-endurance stations, and interoperable data systems. Some participants described how committing major assets early can catalyze community planning and attract complementary projects and open standards for platforms and data can help multiple disciplines couse networks.

Community Partnerships

Kliskey's examples of CBONs resonated with several comments about coproduction, rapid return of usable information to communities, and long-term funding mechanisms that do not lapse between grants. Several participants noted that bidirectional design—asking what research questions local partners want answered and making outputs legible to those users—can make observing more durable.

INTERACTIVE ACTIVITIES: HUMAN CAPACITY AND COLLABORATION

Session 4 concluded with an interactive activity in the style of the World Café method for those in person and breakout discussions for those who participated in the workshop virtually. This section integrates brief report-outs from both activities. Attributions indicate the reporter for each group; individual points reflect what they said on behalf of their groups.

Question 1: Strengthening Science Through Broader Engagement

Allison Miller, Schmidt Ocean Institute, described a broad initial response—"everything." The discussions then narrowed to a set of themes where cocreated work with communities could directly improve science and its utility: coastal and terrestrial systems where people live (e.g., water quality, sea-level rise, extreme events), sea ice and species distributions (including invasive species), pollution and toxin monitoring, and upper atmosphere/ionospheric phenomena. Some participants also mentioned topics that can engage contributors beyond the Arctic (e.g., image classification, genetic sorting, transcribing ship logs). Sea-level rise was discussed as a key issue where polar research connects clearly to global impacts. Miller noted repeated emphasis across discussions on approaches that invite communities to help frame questions and on producing data products that are accessible and useful locally.

Matthew Druckenmiller, National Snow and Ice Data Center, relayed that his group returned often to early and sustained engagement with Indigenous communities—beginning during IPY5 planning—and transparent, open communications about the IPY5 process itself. They discussed engaging youth (roughly ages 5–19) and using multiple languages and suggested a "local-to-global" framing that connects polar science to widely shared concerns. The group also pointed to including the "third pole"—high mountain regions, such as the Himalayas—in IPY5 linkages, drawing on the long history of international organizations engaged in IPY-type efforts, and aligning open-science practices with data sovereignty. Public–private partnerships, storytelling, and collaboration with artists were cited as practical avenues for wider participation and visibility (e.g., social media content that travels).

Question 2: Community-Led Science Beyond Engagement and Empowering Local Voices

Robin Bell, Columbia University, reported that her discussions took stock of current practice. They referenced structured models from Canada, such as territorial research licensing and community-engagement requirements (e.g., Nunavut Research Institute, Aurora Research Institute in the Northwest Territories, Yukon Scientists and Explorers Act), and other examples, such as the study "My Deep Sea, My Backyard" (Amon et al. 2022), a capacity-building initiative that piloted low-cost, locally operated deep-ocean observing (e.g., drop-cams, telepresence, training, codesign) in small island developing states (e.g., Trinidad and Tobago, Kiribati). They noted practical challenges: communities and designated gatekeepers are often overwhelmed; funding pathways can be confusing; and even terminology (e.g., "residents of the north" versus explicit reference to Indigenous peoples) warrants care. Bell said that the group emphasized bringing Indigenous and local partners directly into IPY5 planning and paying them, clarifying proposal processes, asking communities what information would be useful, ensuring results return in usable formats/units, and building local capacity to observe, build, and use tools as a pathway to careers.

Victoria Herrmann, Arctic Institute, said that her group distinguished between Indigenous-led and community-led science, noting overlap but also complementary needs. A recurring theme was sustained codevelopment before, during, and after IPY5, rather than short-term consultation. That implies funding streams that enable Indigenous-led science in place (e.g., Arctic) and community-led efforts in gateway and coastal locations (e.g., Antarctic), and a plural knowledge approach that weaves Indigenous knowledge with academic science throughout project life cycles. The group also discussed cultivating the next generation of leaders in place (e.g., middle and high school students who will be researchers during IPY5) and connecting IPY4 to IPY5 in ways that build continuity. They suggested moving from knowledge creation alone to codeveloping systems and practices—for example, exploring plural knowledge approaches in naming or mapping frameworks.

Question 3: Role of International Collaboration in Fostering the Next Generation of U.S. Polar Researchers and Mechanisms to Facilitate Career Development and Training

Bill Muntean, Center for Strategic and International Studies, summarized three themes. First, leverage organizations rather than creating new ones: SCAR, International Arctic Science Committee (IASC), APECS, and professional societies, such as the American Geophysical Union and European Geophysical Union, already offer fellowships, working groups, and convening power; a practical addition could be a "clearinghouse" that matches ECRs to international opportunities and collaborators. Second, place ECRs in leadership roles and facilitate knowledge transfer from senior experts, including by treating IPY4 outputs as a foundation. Third, use existing codes and frameworks (e.g., Inuit/Greenlandic guidance, society codes of conduct), and embed local partners within the global polar research ecosystem.

Trista Vicks-Majors, Michigan Technological University, said that her group underscored the value of international collaboration for scientific breadth and career development, especially amid current U.S. funding uncertainties. Mentors can introduce ECRs to global networks; in some cases, moving internationally is a viable pathway when positions or resources are constrained at home. Echoing Muntean's "clearinghouse" strategy, her group cited IASC, SCAR, and APECS resources and suggested compiling a shared list of fellowships, exchanges, working groups, and training opportunities in one place to improve access.

Question 4: Recent Campaigns, Additional Capacities, and Skills

Britney Schmidt, planning committee member from Cornell University, drew crosscutting threads: retention—"keeping people in the field," broadly construed—benefits from soft-skills training (e.g., team management; safe, inclusive field practices), a community clearinghouse for lessons learned, and attention to "tail-end" funding so that data can be processed and shared. She said that some participants spoke to lowering barriers for instrumentation and engineering partnerships, sustaining technical and logistics expertise, and making physical qualification processes more inclusive. One concrete idea was a "program inclusion/endorsement" framework: projects funded in the IPY5 window could opt in to shared expectations (e.g., conduct, minimum data practices, common goals), which may broaden visibility and access for non-PI contributors. Schmidt also relayed ideas to improve logistics: flexible access to polar infrastructure across nations; pooling/sharing seats, berths, and field capacity; routine postfield debriefs that feed back into program reviews; and closer integration of scientists into planning to avoid decisions that create downstream complications.

Kristina Collins, Space Sciences Institute, divided her group's comments into two parts. On lessons learned, she said that conduct and reporting practices were a central concern in international and multidisciplinary settings. While cultural differences exist, the discussion pointed to harmonizing codes of conduct in advance, recognizing power dynamics, and ensuring mechanisms (e.g., ombuds roles) that enable reporting and accountability. She cautioned that processes can, at times, be misused to avoid accountability; designing systems with that risk in mind was viewed as important. On capacities and skill sets, she noted that "early career" spans diverse situations (e.g., students, multiple models of postdoctoral roles, people entering and exiting academia) and that transparent, ongoing coordination structures can help ECRs find, contribute to, and grow within international collaborations. Maximizing field opportunities for skill development and supporting community-facing coordinators were mentioned as practical steps.

8

Moving the Science Forward for IPY5

Jamin Greenbaum, planning committee member from the Scripps Institution of Oceanography, moderated the final panel. Greenbaum underscored the necessity and the difficulty of international work in the polar regions. Polar risks and societal vulnerabilities are interconnected, and while the polar community has a long history of collaboration (e.g., the Antarctic Treaty, multinational Arctic efforts), practical barriers persist: funding cycles that do not align across countries; a strong reliance on informal networks; and limited "structured facilitation," especially for early-career researchers (ECRs). He pointed to existing instruments, such as lead agency agreements, and invited panelists to share concrete models, cautionary tales, and ideas that could lower barriers and expand access as the International Polar Year (IPY) 5 takes shape. Panelists were provided with the following discussion questions:

- How do we develop internationally coordinated research questions and partnerships, from idea development, funding, to operations/implementation?
- Are there new/emerging opportunities to promote and implement this coordination (e.g., remote and/or share technologies)?
- How can we overcome known challenges (e.g., formal bilateral national agreements)?
- What role might overarching nongovernmental organizations (e.g., the World Meteorological Organization, Intergovernmental Ocean Commission, Global Ocean Observing System) play?

- What is the role of international committees (e.g., International Arctic Science Committee and Scientific Committee on Antarctic Research) and ongoing priority-setting activities (e.g., Fourth International Conference on Arctic Research Planning)?

BUILDING COLLABORATION MODELS FROM IPY4 TO TODAY

Julia Wellner, University of Houston, organized her remarks around "how" international programs are structured and executed, drawing lessons from three models. First, she described the IPY4 LARsen Ice Shelf System Antarctica (LARISSA) effort on the Antarctic Peninsula. Three National Science Foundation (NSF)-funded U.S. projects—spanning marine and terrestrial geology, glaciology and oceanography, and marine microbiology—were expanded through partnership with international logistics, specifically a connection with the Korea Polar Research Institute (KOPRI). That sequencing mattered: U.S. science awards came first, and international partners then amplified scope and capability. She noted the personal dimension as well: as a postdoc invited to join as a co-principal investigator (PI), she found LARISSA career defining and a reminder to intentionally include ECRs in leadership roles.

Her second example, the International Thwaites Glacier Collaboration (ITGC), operated as a coordinated suite of eight U.S.–U.K. projects with a dedicated coordination office and an explicit ~50/50 funding split between the two lead agencies (NSF and the U.K. Research and Innovation/Natural Environment Research Council). Wellner focused on Thwaites Offshore Research to show how additional partners from Sweden and Germany enlarged the program's reach with people, funds, and field assets. While acknowledging management challenges in parts of the larger portfolio, she emphasized that marine components performed efficiently and delivered critical observations. Third, she highlighted lead agency agreements as a flexible pathway for bottom-up, PI-led teaming when a formal bilateral program does not exist, and she urged U.S. groups to use SCAR processes to keep efforts synchronized with ongoing international planning. She closed by stressing that the ice–ocean interface (e.g., glacier termini, ice-shelf fronts, grounding zones) is changing on seasonal-to-annual timescales. Archives and legacy datasets are helpful but not sufficient. To get credible sea-level projections, she said, teams need to go back repeatedly to those margins and make new, direct measurements (e.g., bathymetry, ocean properties, melt rates, damage state).

LEVERAGING TOURISM FOR SCIENCE AND AMBASSADORSHIP

Allison Cusick, Scripps Institution of Oceanography, presented FjordPhyto, a citizen science program that partners with Antarctic tour operators to sample phytoplankton and meltwater influences along the Antarctic Peninsula. Since 2016, the project has worked with 16 operators and >24 vessels, training expedition staff and engaging >10,000 travelers to collect seasonal data (November–March) that complements station-based and Long-Term Ecological Research efforts. The approach uses repeated cruise tracks to build time series across many fjords and months that are otherwise difficult for national programs with limited ship time to achieve. Social science surveys of participants indicate strong educational impact and durable "Antarctic ambassadorship," with travelers sharing their experiences when they return home.

Cusick highlighted a diverse funding mosaic that made early growth possible—NSF Office of Polar Programs and National Aeronautics and Space Administration (NASA) citizen science programs, the small NASA Science Mission Directorate Community of Practice for Education that supports ECR-oriented seed grants, SCAR fellowships, and operator in-kind support. She argued that IPY5 can scale this model to broaden sampling and public engagement simultaneously with structured training, standardized kits, and short fellowships that place students, scientists, and educators with tour operator partners. Champions are likely important to knit the pieces together, but the opportunity is to increase both scientific coverage and societal reach without overburdening national fleets.

SHIPS OF OPPORTUNITY, REAL-TIME DATA, AND NATIONAL STRUCTURES

Andrew Luke King, Norwegian Institute for Water Research, drew on experience in both United States and European systems. King focused on the Norwegian Ships of Opportunity Program (NorSOOP) that equips ships with suites of sensors—FerryBoxes that make seawater plus meteorological and radiometric measurements—to sail regular routes in the Arctic and sub-Arctic. The program supports these platforms, which deliver essential ocean variables with near real-time satellite uplink and serve as calibration/validation (cal/val) for remote sensing products. Touchscreen displays aboard passenger vessels simultaneously enable ocean literacy and citizen science engagement. Similar instrumentation is operating on Canadian Arctic transits and some ships crossing to the Antarctic Peninsula, building richly sampled transects and supporting process studies (e.g., land–ocean connectivity, sea ice melt effects on productivity).

King then mapped the funding and coordination landscape: the Research Council of Norway funds basic science and long-term infrastructures (e.g., Nor-

SOOP); Integrated Carbon Observation System provides funding for land- and ocean-based carbon dioxide stations; Institute of Marine Research and Norwegian Polar Institute operate regional research fleets; and Horizon Europe and the European Strategy Forum on Research Infrastructures supply sustained frameworks for multicountry projects. Polar research is coordinated at various EU levels through the EuroGOOS Arctic Regional Ocean Observing System (ROOS) and Arctic Ocean Alliance, European Polar Board, EU PolarNet, and EU Polar Cluster. He emphasized mobility pathways (e.g., Marie Skłodowska-Curie Actions, ECR Starting Grants, transnational "access" programs through the Aqua Research Infrastructure Services and Arctic Research Icebreaker Consortium EU projects) that ECRs outside Europe can also pursue. Lessons for IPY5 include leveraging routine routes and platforms for repeat observations, using regional alliances, such as the Global Ocean Observing System (GOOS) and Arctic ROOS to align methods and data, and planning funding with continuity in mind so that observing lines persist beyond a 2-year IPY window.

FROM SHARED QUESTIONS TO SHARED TOOLS

Won Sang Lee, the Korea Polar Research Institute, addressed two practical questions: how to develop internationally coordinated science priorities and equip teams to execute them. He noted that horizon-scanning exercises (e.g., within SCAR) offer a starting list that research communities refine into focused programs. Different models then exist to deliver those priorities: bilateral programs (e.g., ITGC) can be highly productive, but multinational constructs may offer resilience when geopolitics are uncertain and requirements (e.g., ships, aircraft, under-ice access) are too large for any two nations. He pointed to ongoing multinational planning for full-season polar campaigns—akin in spirit to the Multidisciplinary Drifting Observatory for the Study of Arctic Climate—as venues where IPY5 could naturally plug in and accelerate progress.

Lee described the Land-Ice/Ocean Network Exploration Using Semiautonomous Systems (LIONESS) (see Figure 8-1) as a collaborative model designed to link land, ice, and ocean observations in service of a central science goal: better constraining the Antarctic Ice Sheet's contribution to future sea-level rise. He situated it within existing international priority-setting and programmatic scaffolding (e.g., SCAR Horizon Scan, Antarctica International Science and Infrastructure for Synchronous Observation) and explicitly tied the panel's discussion prompts to practical coordination needs—from codefining questions and funding pathways to operations—and the roles for the World Meteorological Association, Intergovernmental Oceanographic Commission, GOOS, and SCAR and International Arctic Science Committee in knitting efforts together.

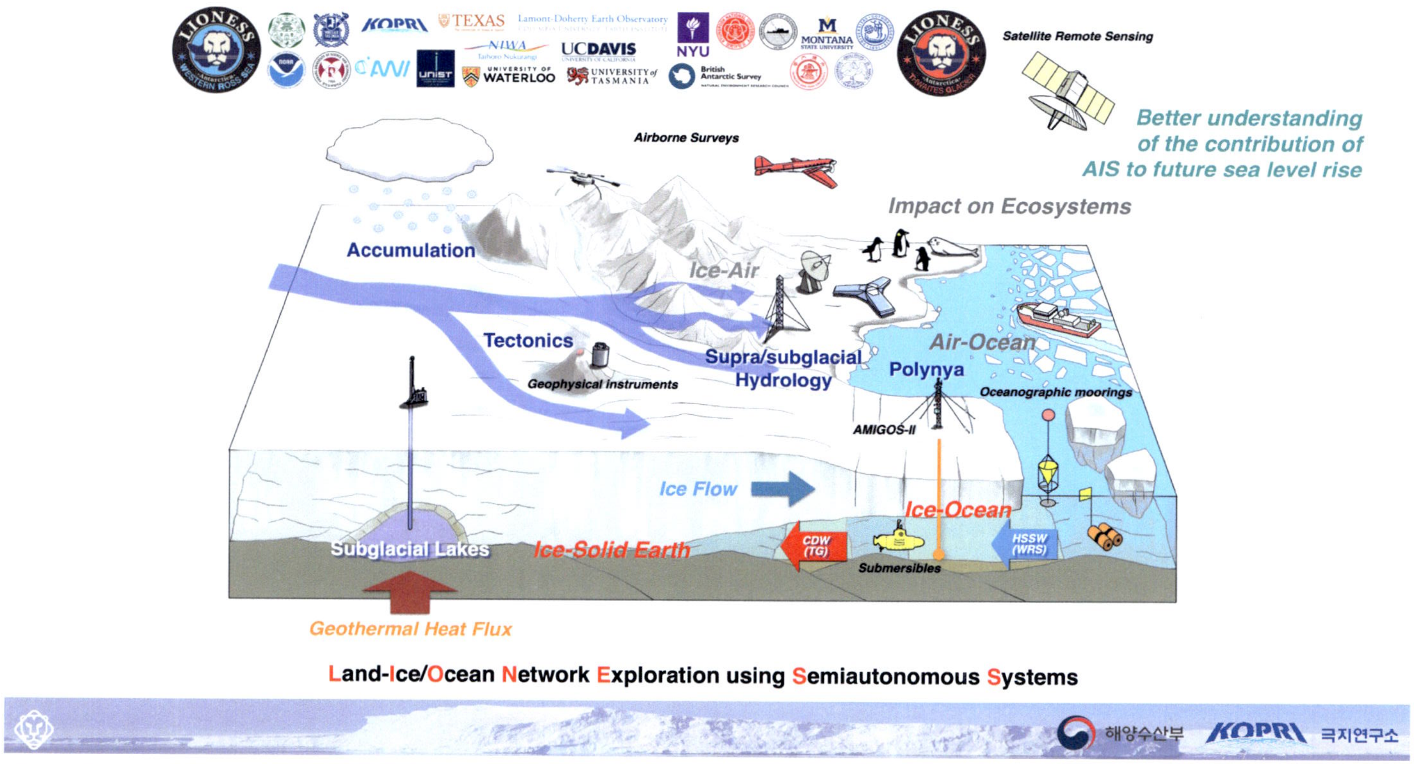

FIGURE 8-1 The Land-Ice/Ocean Network Exploration Using Semiautonomous Systems (LIONESS) collaborative model.
NOTE: The collaborative links land, ice, and ocean observations for a better understanding of the contribution of the Antarctic Ice Sheet (AIS) to future sea-level rise.
SOURCE: Presentation by Won Sang Lee on May 22, 2025; design by Won Sang Lee.

Lee also emphasized technology development for the "mission impossible" parts of observation, especially under-ice-shelf access and sustained measurements in harsh environments. Lee noted that the ice–ocean–atmosphere system is intrinsically coupled and complex, requiring multiplatform observing across interfaces (ice–ocean, ice–air, air–ocean) using airborne surveys, satellites, geophysical instruments, oceanographic moorings, AMIGOS-II nodes, submersibles, etc. He illustrated this with a national program, the Korea Network for Observation and Prediction of Ice Sheet and Sea-Level Changes in a Warming World (2023–2031)—and LIONESS itself, which incorporates those elements into an integrated, semiautonomous network focused on key processes (e.g., subglacial hydrology, geothermal flux, polynyas, circumpolar deep water/high-salinity shelf water exchange) that govern ice–ocean coupling.

Lee cited top-level bilateral dialogues (e.g., between Korean and U.S. science ministries/offices) that can seed joint programs focused on complex Earth system interactions and sea-level drivers. Finally, he described KOPRI's support for ECR networks and mentorship, arguing that ECRs could help set priorities and gain hands-on experience with new platforms; institutions can play a formal role in provisioning mentorship, convening, and "small grant" glue.

INTERNATIONAL COLLABORATIVE EXPLORATION OF THE CRYOSPHERE THROUGH AIRBORNE PROFILING (ICECAP) LESSONS: FUNDING FLEXIBILITY, FIELD ROLES, AND DATA LEGACY

Jason Roberts, Australian Antarctic Division, discussed the ICECAP program, launched during IPY4 and sustained through a sequence of NSF, NASA (Operation IceBridge), the European Space Agency cal/val activities, and private foundation support. From hubs across East Antarctica, small field teams flew repeated seasons to map bed topography, bathymetry, gravity, and magnetics—datasets that have greatly improved ice sheet and ocean modeling. He traced a "potted history" of funding that evolved from ice sheet characterization to grounding line and ocean frontier focus (including airborne oceanography) and to satellite cal/val (e.g., the European Space Agency's satellite mission: Biomass Monitoring Mission for Carbon Assessment) showing how the science agenda and instrument suite coevolved.

The operational lesson was stark, he said: building a multiagency, multicountry program is like balancing a "house of cards on a shaker table." National reviews run on different clocks; mutual dependencies cannot be assumed in proposals; and discretionary funds have often been the bridge that prevented collapse. He argued for robust lead agency style instruments to decouple funding risks, clear preseason science priorities to avoid infield friction, and formal leadership roles for ECRs, even in small teams—so that career records reflect actual responsibility. ICECAP's enduring value, he noted, lies in both its data legacy and the muscle memory it provides for organizing complex international field programs.

DISCUSSION

Questions from the room and virtual Q&A platform revisited structural hurdles and possible IPY5 instruments.

Timing Mismatches

Some panelists noted that misaligned national funding clocks can jeopardize otherwise strong partnerships; the United States can be slower than some counterparts. Discretionary bridge funds and staggered coordination help, but only partially. From the EU side, Horizon Europe can stand up large, multicountry projects quickly; however, 4–5-year durations are still short for sustained observing, so continuity plans are beneficial.

Co-funding Pitfalls

Several speakers noted that "50/50" balance sounds straightforward until in-kind contributions (e.g., ship time, aircraft, logistics) must be valued. Past efforts have failed at this hurdle. Some participants suggested articulating partner-specific benefits early, agreeing on valuation frameworks up front, and anticipating overhead practices that can distort budgets.

Coordination Venues

SCAR was highlighted as a grassroots incubator for internationally coordinated questions and networks, with travel support, disciplinary groups, and regular meetings. For logistics matchmaking, the Council of Managers of National Antarctic Programs can help identify capacity (e.g., empty berths, aligned routes). Several panelists also recognized a frequent reliance on personal networks; an IPY5 "clearinghouse" for partners and needs could lower that barrier.

Lead Agency Agreements for IPY5

In closing, Greenbaum proposed exploring a bespoke, timebound "IPY5 lead agency" instrument—an umbrella under which national agencies commit to synchronized calls and adjudication, common expectations (e.g., ECRs on both sides of each proposal), and shared standards (e.g., open data, field conduct expectations). This would be ambitious, but many panelists viewed it as aligned with the scale of IPY5.

Overall, the discussion emphasized that the form of collaboration—funding instrument, coordination office, clearinghouse, logistics framework—is as important as the content (e.g., science questions, methods). IPY5 planning offers a window to build both, in ways that broaden participation and make the next generation of international polar research more durable.

9

Forming the Vision for the Future—Takeaways and Examples of Potential Next Steps

The workshop's final activity, moderated by Craig Lee, was an opportunity for any participants to provide brief comments on their key takeaways and what they viewed as opportunities to take the research questions and move toward action and scientific advancement for International Polar Year (IPY) 5. The following sections summarize what they emphasized.

Look Back to Move Forward

Allison Cusick asked about a formal synthesis of what worked (and did not) in IPY4 and suggested using recent "field reports" and the social media landscape to update those lessons. Igor Krupnik, Smithsonian Institute, reinforced that much of IPY4's scaffolding is still relevant: a planning report (NRC, 2004), a joint committee that required international participation, and active legacy organizations, such as the Association of Polar Early Career Scientists (APECS). He urged IPY5 to mine that near-term institutional memory rather than reinvent it.

How to Plug In

An audience member asked for concrete next steps and pathways for early-career researcher (ECR) involvement. Lee outlined a few near-term tracks: bottom-up, community planning in scientific venues (e.g., Scientific Committee on Antarctic Research, International Arctic Science Committee annual Arctic Science Summit Week conferences) even without a designated U.S. lead agency and additional activities to extend and coordinate planning.

On datasets at risk (e.g., decommissioning), Jamin Greenbaum noted an active, multi-institution effort led by Steve Diggs at University of California—San Diego that is inventorying and coordinating preservation.

Early-Career Researchers and Indigenous Partners

Russ Shamberg, University of Maryland, offered that ECRs can help organize sessions and committees as a way to collaborate with senior scientists and share knowledge. APECS representatives invited ECRs to submit priorities to the Polar Early Career World Summit synthesis; it sits on the international IPY5 working group on capacity building. Greenbaum suggested writing ECR inclusion into new IPY5 lead agency agreements (e.g., requiring ECR leadership on both international proposal teams). He cited National Aeronautics and Space Administration (NASA) practices (e.g., double-blind peer review, ECRs on leadership teams) as models that also cut through institutional rules that prevent postdocs from serving as co-PIs.

A participant suggested that Indigenous scientists and governance leaders are distinct from ECRs and could be explicitly invited into planning and decision roles if IPY5 is to "grow beyond research."

A Blueprint from IPY 4 Experience

Robin Bell, Lamont-Doherty Earth Observatory at Columbia University, offered an example blueprint drawn from her IPY4 experience:

- Resource the planning: Explore funding opportunities now to support U.S. IPY5 planning and ensure U.S. leadership internationally.
- Support participation: Fund ECRs, Indigenous partners, and community leaders to attend planning meetings and build collaborations that seed future program leadership.
- Think ahead: Build the plan now, so that it is ready when agency engagement accelerates.
- Explain why polar science matters: Sharpen public and Congressional messaging (e.g., economics, security, forecasts, risk management) so that IPY5 is poised to be impactful, international, and novel.
- Be bold: Build human networks first, then observational networks; consider "radical" scope (e.g., include the third pole or even "everywhere it snows") to connect poles with people's daily lives.

Additional Themes

Some participants stressed that IPY5 can be used to restore and grow programs. Britney Schmidt argued that clear, ambitious "mission-like" goals help win support (as NASA missions do) and rebuild capacity. Multiple speakers stressed coordinated storytelling that scales beyond "100 one-offs," harnessing social media and professional communicators to reduce burden on communities and scientists. Several participants suggested central inventories, guidance, and funding to preserve vulnerable data assets and make data products findable and usable by nonscientist communities. Other participants noted the importance of "year zero" funding (i.e., relationship building before research), tail-end support to analyze/publish/curate data, and shared recipes for codes of conduct, logistics, and safety across international teams.

Closing Remarks

Planning committee chair, Craig Lee, closed by acknowledging the importance of Indigenous participation and committing to more venues that bring Indigenous and local leaders directly into planning. He emphasized that, while structural fixes (e.g., funding mechanisms, lead agency agreements, codes of conduct) may be essential, science advances through people, such as senior scientists who can invite, mentor, and bring students and postdocs into the room and share responsibility and visibility; and students and ECRs who can speak up and step up to take responsibilities when offered and help organize and connect communities.

References

Alley, Richard B., Sridhar Anandakrishnan, Knut Christianson, Huw J. Horgan, Atsu Muto, Byron R. Parizek, David Pollard, and Ryan T. Walker. 2015. "Oceanic Forcing of Ice-Sheet Retreat: West Antarctica and More." Annual Review of Earth and Planetary Sciences 43 (1): 207–31. https://doi.org/10.1146/annurev-earth-060614-105344.

Amon, Diva J., Randi D. Rotjan, Brian R. C. Kennedy, Gerard Alleng, Rafael Anta, Eriatera Aram, and Thera Edwards, et al. 2022. "My Deep Sea, My Backyard: A Pilot Study to Build Capacity for Global Deep-Ocean Exploration and Research." Philosophical Transactions of the Royal Society B: Biological Sciences 377 (1854): 20210121. https://doi.org/10.1098/rstb.2021.0121.

Bamber, J. L., A. J. Tedstone, M. D. King, I. M. Howat, E. M. Enderlin, M. R. van den Broeke, and B. Noel. 2018. "Land Ice Freshwater Budget of the Arctic and North Atlantic Oceans: 1. Data, Methods, and Results." Journal of Geophysical Research: Oceans 123 (3): 1827–1837. https://doi.org/10.1002/2017JC013605.

Bronselaer, Ben, Michael Winton, Stephen M. Griffies, William J. Hurlin, Keith B. Rodgers, Olga V. Sergienko, Ronald J. Stouffer, and John L. Russell. 2018. "Change in Future Climate Due to Antarctic Meltwater." Nature 564 (7734): 53–58. https://doi.org/10.1038/s41586-018-0712-z.

Canadell, Josep G., Paulo M. S. Monteiro, Marcos H. Costa, Leandro Cotrim da Cunha, Peter M. Cox, Andrey V. Eliseev, Stephanie Henson, et al. 2021. "Global Carbon and Other Biogeochemical Cycles and Feedbacks." In Climate Change 2021: The Physical Science Basis, edited by Valérie Masson-Delmotte et al., 673–816. Cambridge, UK, and New York: Cambridge University Press. https://www.cambridge.org/core/books/climate-change-2021-the-physical-science-basis/global-carbon-and-other-biogeochemical-cycles-and-feedbacks/93DFD13E855AC1F1B502965CABE28B7F.

Falvo, Grant, Edward A. G. Schuur, Eugénie S. Euskirchen, Susan M. Natali, Oliver Sonnentag, Haley Alcock, Kyle Arndt, et al. 2025. "Record 2024 Winter Carbon Emissions Coincide with Record Warmth across Boreal Forest, Tundra, and Wetland Ecosystems." Environmental Research Letters 20 (11). https://doi.org/10.1088/1748-9326/ae09bb.

Herrmann, A. 2024. Landscape of Tribal Communities: Context for Working in Rural Alaska. Fairbanks: Alaska Center for Climate Assessment and Policy, University of Alaska Fairbanks. https://uaf-accap.org/wp-content/uploads/2024/06/Landscape-of-Tribal-communities.pdf.

LaRue, Michelle A., David T. Iles, Stéphane Labrousse, Peter Fretwell, Daniel Ortega, Emma Devane, Isabel Horstmann, et al. 2024. "Advances in Remote Sensing of Emperor Penguins: First Multi-Year Time Series Documenting Trends in the Global Population." Proceedings of the Royal Society B 291 (2018): 20232067. https://doi.org/10.1098/rspb.2023.2067.

Lee, Craig M., Michael DeGrandpre, John Guthrie, Victoria Hill, Ron Kwok, James Morison, Christopher J. Cox, Hanumant Singh, Timothy P. Stanton, and Jeremy Wilkinson. 2022. "Emerging Technologies and Approaches for In Situ, Autonomous Observing in the Arctic." Oceanography 35 (3–4): 210–221. https://doi.org/10.5670/oceanog.2022.127.

Markon, Carl, Sarah Gray, Matthew Berman, Leticia Eerkes-Medrano, Thomas Hennessy, Henry Huntington, Jeremy Littell, et al. 2018. "Alaska." In Impacts, Risks, and Adaptation in the United States: Fourth National Climate Assessment, Volume II, edited by David R. Reidmiller et al., 1185–1241. Washington, DC: U.S. Global Change Research Program. https://pubs.usgs.gov/publication/70201870.

National Research Council (NRC). 2004. A Vision for the International Polar Year 2007–2008. Washington, DC: National Academies Press. https://doi.org/10.17226/11013.

Orr, Thomas, Carl Cameron, Heidi Dietterich, Meghan Loewen, Thomas Lopez, Jamie Lyons, Janet Nakai, et al. 2024. 2020 Volcanic Activity in Alaska—Summary of Events and Response of the Alaska Volcano Observatory. Scientific Investigations Report 2024–5004. Reston, VA: U.S. Geological Survey. https://doi.org/10.3133/sir20245004.

Painter, Scott L., Ethan T. Coon, Ajay J. Khattak, and Julie D. Jastrow. 2023. "Drying of Tundra Landscapes Will Limit Subsidence-Induced Acceleration of Permafrost Thaw." Proceedings of the National Academy of Sciences 120 (8): e2212171120. https://doi.org/10.1073/pnas.2212171120.

Rantanen, Mika, Alexey Yu. Karpechko, Antti Lipponen, Kalle Nordling, Otto Hyvärinen, Kirsi Ruosteenoja, Timo Vihma, and Ari Laaksonen. 2022. "The Arctic Has Warmed Nearly Four Times Faster Than the Globe Since 1979." Communications Earth & Environment 3: 168. https://doi.org/10.1038/s43247-022-00498-3.

Rignot, Eric. 2023. "Observations of Grounding Zones Are the Missing Key to Understand Ice Melt in Antarctica." Nature Climate Change 13: 1010–13. https://doi.org/10.1038/s41558-023-01819-w.

See, Craig R., Anna-Maria Virkkala, Susan M. Natali, Brendan M. Rogers, Marguerite Mauritz, Christina Biasi, Stef Bokhorst, et al. 2024. "Decadal Increases in Carbon Uptake Offset by Respiratory Losses across Northern Permafrost Ecosystems." Nature Climate Change 14. https://doi.org/10.1038/s41558-024-02057-4.

Seroussi, Hélène, Antony J. Payne, Frank Pattyn, Sophie Nowicki, Pierre Mathiot, Mariano Montoya, Mathieu Morlighem, et al. 2024. "Evolution of the Antarctic Ice Sheet over the Next Three Centuries from an ISMIP6 Model Ensemble." Earth's Future 12 (9): e2024EF004561. https://doi.org/10.1029/2024EF004561.

Stroeve, Julienne, Adam Crawford, Steven Ferguson, Ian Stirling, Lauren Archer, Graeme York, David Babb, and Randall Mallett. 2024. "Ice-Free Period Too Long for Southern and Western Hudson Bay Polar Bear Populations If Global Warming Exceeds 1.6 to 2.6°C." Communications Earth & Environment 5: 296. https://doi.org/10.1038/s43247-024-01430-7.

Stroeve, Julienne C., Dirk Notz, Jacqueline Dawson, Edward A. G. Schuur, Dorthe Dahl-Jensen, and Christian Giesse. 2025. "Disappearing Landscapes: The Arctic at +2.7°C Global Warming." Science 387 (6734): 616–21. https://doi.org/10.1126/science.ads1549.

Sutton, Olivia F., Christopher P. R. McCarter, and J. Michael Waddington. 2024. "Globally Significant Arsenic Release by Wildfires in a Mining-Impacted Boreal Landscape." Environmental Research Letters 19 (6). https://doi.org/10.1088/1748-9326/ad461a.

Virkkala, Anna-Maria, Brendan M. Rogers, Jennifer D. Watts, Kyle A. Arndt, Stefano Potter, Isabel Wargowsky, Edward A. G. Schuur, et al. 2025. "Wildfires Offset the Increasing but Spatially Heterogeneous Arctic–Boreal CO_2 Uptake." Nature Climate Change 15 (2): 188–95. https://doi.org/10.1038/s41558-024-02234-5.

Vogt, Judith, Martijn M. T. A. Pallandt, Luana S. Basso, Abdullah Bolek, Kseniia Ivanova, Mark Schlutow, Gerardo Celis, et al. 2025. "ARGO: Arctic Greenhouse Gas Observation Metadata Version 1." Earth System Science Data 17 (6): 2553–73. https://doi.org/10.5194/essd-17-2553-2025.

A

Workshop Agenda

COMMITTEE ON EXPLORING KEY RESEARCH AND MONITORING TOPICS FOR U.S. ENGAGEMENT IN THE FIFTH INTERNATIONAL POLAR YEAR—A WORKSHOP

Carnegie Institute for Science
5251 Broad Branch Rd NW, Washington D.C., 20015
MAY 20, 2025
GREENEWALT AUDITORIUM

10:00–10:15 Welcome and Introduction
Craig Lee, *Planning Committee Chair*, University of Washington

SESSION 1
POLAR SCIENCE FOR SOCIETY

How does Arctic environmental change impact the lives of northern residents, and how do changes in both poles affect lives at lower latitudes? Polar research provides understanding and data products to inform decisions across a wide range of sectors, including human health (e.g., predictions of harmful algal blooms, vectors of infectious disease), zoning and property insurance in coastal regions (through refined estimates of sea-level rise), predictions of communications disruptions (by providing data to constrain space weather models), and understanding how ecosystem shifts impact both commercial and subsistence hunting and fishing, among many others. This session will provide an overview of some of the ways in which society is impacted by polar sciences.

10:15–10:45 Overview of the Polar Regions
Richard Alley, *Keynote Speaker*, The Pennsylvania State University

10:45–11:35 **Panel 1: Applications of Polar Science**

Andrew Lloyd, *Moderator, Planning Committee Member*, Lamont-Doherty Earth Observatory/Columbia University
Robert McCoy, *Speaker*, University of Alaska—Fairbanks
Alisa Alexander, *Speaker*, University of Alaska—Fairbanks
Jeremy Wilkinson, *Speaker*, British Antarctic Survey

Discussion Questions

- What data/information do you rely on in your position? How do you use this data/information?
- Are there any areas/issues that impact where IPY5 might offer an opportunity to fuel a significant advance in understanding or data delivery that could ease your work and/or improve the quality of the services you provide?

SESSION 2
SCIENTIFIC QUESTIONS AND SCIENCE TO MEET MULTIDOMAIN OPERATIONS NEEDS

Understanding of the polar regions, their role in the broader Earth system, and linkages to space, have advanced through decades of research and prior IPYs, yet questions remain. As we approach 2032–33, the polar regions are experiencing rapid change, with severe impacts to local ecosystems and communities and global implications in areas that include sea-level rise, weather, food security, shipping and societal well-being. Refined understanding of these changes can provide improved predictions to help society plan for and adapt to the future. This session sets the stage for workshop focus centered on identifying and exploring ambitious research goals that could leverage the span and scope of IPY5 to achieve transformative advancements in understanding of the polar regions, their role in the broader Earth system, and the impacts of polar environmental change on society. Discussions in this section and throughout the meeting will aim to identify "big-picture" problems than span multiple disciplines, as well as large, multidomain operational needs, such as observing systems or forecast centers, meant to inform decision making across a range of spatial and temporal scales.

11:35–12:45 **Panel 2: Key Scientific Questions and Science to Meet for Multidomain Operations**

Scott Goetz, *Moderator, Planning Committee Member*, Northern Arizona University
Jacqueline Austermann, *Speaker*, Columbia University
Allison Jaynes, *Speaker*, University of Iowa
Ted Schuur, *Speaker*, Northern Arizona University
Sheyna Wisdom, *Speaker*, Alaska Ocean Observing System
Jacqueline Grebmeier, *Speaker*, University of Maryland
Michelle LaRue, *Speaker*, University of Canterbury

Discussion Questions

- What aspects of these example areas can be uniquely explored during IPY5? What are the roles of in situ observing, remote sensing and numerical modeling in advancing understanding in these areas?
- What research areas could meet multidomain operational needs (i.e., have multiple uses across research disciplines and/or applications)?
- What technologies and capabilities are currently available, or expected to be available in 2032, to enable addressing these questions? What opportunities might exist for international collaboration and coordination to facilitate addressing these questions?
- Are there other specific needs/foci within these broad examples that should be brought to the forefront of our thinking?
- Question to participants: what other key topics could be explored during IPY5, and what operational needs might they address?

12:45–1:45 Networking Lunch and Time to Contribute to Key Questions "Wall"

1:45–2:30 **Session 2 Interactive Activity: Scientific Questions and Science for Multidomain Operations Needs**

Discussion Questions

- What critical research questions do you think should be considered for IPY5? These could build on examples discussed in the talks or be different. What societal implications/interests might these research questions inform? What types of experts/groups would be involved in addressing these questions?
- What bipolar research questions can best be addressed through an approach that links the Arctic and Antarctic? What societal implications/interests would this inform? What types of experts/groups would be involved in addressing these questions?
- What environmental services/operational needs might an IPY5-driven effort aim to address? What structures would need to be built to provide those services? What types of experts/groups would be involved in addressing these questions?
- Are there different research questions that could be addressed synergistically, utilizing the same/similar infrastructure or datasets? What might a collaboration of this type entail?

2:30–3:00	Interactive Activity Plenary Report Out **Craig Lee,** *Planning Committee Chair,* University of Washington
3:00–3:15	Networking Break

SESSION 3
APPROACHES AND CAPABILITIES TO SUPPORT IPY5

The remote, harsh environment of the polar regions presents unique challenges for observational research, complicating the task of collecting sustained measurements of the necessary spatial and temporal scales and scope. Polar research often requires highly specialized equipment and infrastructure (e.g., ice-strengthened vessels, ice core drilling rigs) built to withstand extremely cold conditions. Extensive, costly logistical capabilities are required to deploy and sustain researchers and equipment in remote locations. Though largely free from the environmental challenges faced by in situ observing, remote sensing and numerical modeling rely on in situ measurements, and are thus constrained by sparse data availability, temporal and spatial resolution challenges, and other issues such as frequent cloud cover and seasonal darkness interfering with satellite-based sensing. Rapid changes in technologies and approaches have occurred since IPY4, with advancements continuing to emerge that may enable new and expanded efforts in field, remotely based, and modeling research for IPY5. In this session, we will discuss how to harness new technologies, capabilities, and approaches for IPY5, and how emerging research goals for IPY5 could, in turn, inform further near-term development of these technologies and approaches.

3:15–4:20 **Panel 3: Approaches and Capabilities**

Britney Schmidt, *Moderator, Planning Committee Member*, Cornell University
Matthew Siegfried, *Speaker*, Colorado School of Mines
Brendan Rogers, *Speaker*, Woodwell Climate Research Center
Esmee Van Wijk, *Speaker*, Commonwealth Scientific and Industrial Research Organization
Charles Koven, *Speaker*, Lawrence Berkeley National Laboratory
Terry Wilson, *Speaker*, Ohio State University

Discussion Questions

- What new approaches, capabilities, and technologies are available (or are anticipated to be available by 2032) that could help facilitate research advancements during IPY5?
- Are there opportunities for new interdisciplinary approaches or synergistic use of capabilities/technologies among disciplines that could be leveraged?
- Are there other capabilities or technologies that do not yet exist that are fundamental to addressing any of the priority research areas discussed? Is there a path to reaching these capabilities/technologies?

4:20–5:00 **Session 3 Interactive Activity: Approaches and Capabilities to Support IPY5**

- What new or emerging technologies or capabilities could be utilized to address the research areas discussed in this session?
- Are there any novel tools or approaches to consider (including from other disciplines/applications not typically used in polar research)?
- What is the role of international collaboration in fostering the development and use of new capabilities and approaches? What types of experts/groups would need to be involved in fostering these relationships?
- Are there any technologies or approaches where researchers may be able to help inform development so as to support polar research/IPY5?

5:00–5:30 Interactive Activity Plenary Report Out
Mike Hartinger, *Moderator, Planning Committee Member*, Space Science Institute

5:30 Adjourn Day 1

MAY 21, 2025
GREENEWALT AUDITORIUM

10:00–10:30 Welcome and Recap of Day 1
Craig Lee, *Planning Committee Chair*, University of Washington
Lynn Talley, University of California—San Diego

SESSION 4
HUMAN CAPACITY AND COLLABORATION

Advancements in our understanding of the polar regions are underpinned by expansive teams that extend well beyond the principal investigator(s)/core team for any given project. This can include community members, tourists, and the general public who contribute to data collection in their communities or aboard cruise ships; coordinators who manage complex logistics for deploying personnel and equipment; pilots, captains, and drivers who transport researchers to remote sites; and engineers and technicians that build and maintain specialized equipment and models, among others. Together, these groups provide the human capacity needed to advance polar sciences and will be essential leaders and contributors to IPY5. Within these teams and groups, collaborations—crossing national, international, disciplinary, and cross-sectoral boundaries—are critical for enabling and supporting polar research. Identifying and strengthening partnerships will be an essential element of a successful IPY5.

This session will consider how polar science is accomplished, highlighting many of the types of individuals and teams that enable and contribute to research. It will also look at some specific examples of how research is conducted and the practicalities/logistics required.

10:30–11:15 **Panel 4: Strengthening Human Capacity**

Lil Alessa, *Moderator, Planning Committee Member*, University of Idaho
Ed Brook, *Speaker*, Oregon State University

Twila Moon, *Speaker*, National Snow and Ice Data Center
Hannah-Marie Garcia-Ladd, *Speaker*, Indigenous Sentinels Network

Discussion Questions

- What do you see as the primary strengths and challenges of supporting polar research today?
- Do you envision supporting IPY5? If yes, what might that look like? Do you expect strengths and challenges to change moving forward to IPY5?

11:15–12:15 **Panel 5: Strengthening Research Capacity**

Mike Hartinger, *Moderator, Planning Committee Member*, Space Institute Center
Matthew Shupe, *Speaker*, University of Colorado—Boulder
Ryan Venturelli, *Speaker*, Colorado School of Mines
Katherine Cariglia, *Speaker*, Massachusetts Institute of Technology
Andrew Kliskey, *Speaker*, University of Idaho

Discussion Questions

- What are the primary strengths and challenges to the current approach to conducting research and supporting early-career researcher development? What lessons can be learned from prior campaigns?
- These activities require many skilled individuals beyond the researchers themselves. Are there key areas where capacity development will be required to support IPY5-level activity? How might we do this?
- What opportunities does the ramp up period to IPY5 provide to strengthen this capacity?
- Are there mechanisms to help sustain the needed capacity over time?

12:15–1:15 Networking Lunch

1:15–1:55 **Session 4 Interactive Activity: Human Capacity and Collaboration**

- Among the research questions discussed during this workshop, what do you see as key areas where broadening community/public engagement in research would strengthen the science?

- How can we foster Indigenous and/or community-led science, beyond public engagement? How do we empower local voices to shape polar research?
- What is the role of international collaboration in fostering the next generation of U.S. polar researchers? Are there mechanisms in place that can help to facilitate career development/training?
- How can the lessons learned from the campaigns discussed (or other campaigns) be addressed to strengthen human capacity and collaboration moving forward? Are there other personnel capacity and/or required skill sets that are needed to support/enable IPY5 that have not yet been discussed?

1:55–2:25 Interactive Activity Plenary Report Out
Jenny Baeseman, *Moderator, Planning Committee Member*, Polar Consultant

2:25–2:55 Networking Break

2:55–4:00 **Panel 6: International Coordination and Collaboration—Moving the Science Forward for IPY5**

Jamin Greenbaum, *Moderator, Planning Committee Member*, Scripps Institution of Oceanography
Julia Wellner, *Speaker*, University of Houston
Allison Cusick, *Speaker*, Scripps Institution of Oceanography
Andrew Luke King, *Speaker*, Norwegian Institute for Water Research
Won Sang Lee, *Speaker*, Korea Polar Research Institute
Jason Roberts, *Speaker*, Australian Antarctic Division

Discussion Questions

- How do we develop internationally coordinated research questions and partnerships, from idea development, funding, to operations/implementation?
- Are there new/emerging opportunities to promote and implement this coordination (e.g., remote and/or share technologies)?
- How can we overcome known challenges (e.g., formal bilateral national agreements)?
- What role might overarching nongovernment organizations (e.g., the World Meteorological Organization, Intergovernmental Ocean Commission, Global Ocean Observing System) play?
- What is the role of international committees (e.g., International Arctic Science Committee and Scientific Committee on Antarctic Research) and

ongoing priority setting activities (e.g., Fourth International Conference on Arctic Research Planning)?

- Question to participants: where do you see opportunities to support enhanced international collaboration and coordination?

4:00–4:30 **Forming the Vision for the Future: Meeting Takeaways and Reflections on Next Steps**

Workshop participants will be invited to provide brief comments on their key takeaways from the meeting and what they view as opportunities to take the research questions discussed and move us toward action and scientific advancement for IPY5. Preference will be given first to early-career researchers interested in sharing their reflections.

Craig Lee, *Planning Committee Chair*, University of Washington

4:30–4:35 Closing Remarks

Craig Lee, *Planning Committee Chair*, University of Washington

4:35 Workshop Adjourns

B

Planning Committee Biographies

Craig Lee, Ph.D., is the assistant director for research and senior principal oceanographer in the Applied Physics Laboratory and professor in the School of Oceanography at the University of Washington. His research focuses on studies of small-scale dynamics, physical–biogeochemical interactions, atmosphere–ice–ocean dynamics, and Arctic change. He leads a small team of scientists and technologists that creates new instruments and approaches to collect novel observations, developed the Seaglider family of long-endurance gliders, and works with a broad range of other platforms. His leadership roles include chairing the University National Oceanographic Laboratory System Council (the advisory body for the U.S. Academic Research Fleet) and Observing Panel for the Study of Environmental Arctic Change; he served as subject-matter expert for the U.S. delegation to the G7 Future of Seas and Oceans Initiative Arctic Group and chief scientist for numerous midscale programs. He received a National Science Foundation (NSF) Creativity Award for research into the impact of small-scale dynamics on biological and biogeochemical processes. Dr. Lee has a B.S. in electrical engineering and computer science from the University of California—Berkeley and a Ph.D. in physical oceanography from the University of Washington, and he did postdoctoral research in the department of Physical Oceanography at the Woods Hole Oceanographic Institution. He was a member of the planning committee for the National Academies of Sciences, Engineering, and Medicine (the National Academies) Workshop on Technology Developments to Advance Antarctic Research.

Lilian Alessa, Ph.D., is a President's Professor and codirector of the Center for Resilient Communities at the University of Idaho and an affiliate faculty member with George Mason University, the University of Alaska—Fairbanks, and Arizona State University. Her work spans the social and physical sciences, and she has over 30 years of experience with academic, federal, state, local, tribal, and territorial partners across Canada and the United States, with an emphasis on Arctic all-domain awareness, climate and environmental change, national security, defense, and the resilience of systems, people, and communities. Dr. Alessa pioneered the development and best practices for community-based observing networks, which leverage place-based and local knowledge to measure environmental change and its context for societal well-being. She has contributed to the global Arctic Observing Network and led the North Atlantic Treaty Organization Multinational Capabilities Development Campaign Arctic adaptation portfolio. She has also served on the U.S. National Climate Assessment Team and board of advisors of the National Defense Industrial Association. She received the Governor General's Award in Canada and multiple academic and service awards. Dr. Alessa has a Ph.D. in cell biology from the University of British Columbia. She was a member of the National Academies Committee on Leveraging the Future Research and Development Ecosystem for the Intelligence Community.

Jenny Baeseman, Ph.D., is a consultant for national and international polar affairs based in Wisconsin. She served as the founding director of the Association of Polar Early Career Scientists, director of the World Climate Research Programme's Climate and Cryosphere Project, and executive director for the Scientific Committee on Antarctic Research (SCAR). Her prior research focused on the interactions between biology, geology, and chemistry, working in extremely contaminated environments to the pristine streams of the McMurdo Dry Valleys in Antarctica. Her current work involves administering and coordinating cryosphere activities, consulting on polar and climate policy, and advising on educational projects. She has been honored as a Distinguished Alumnus from the University of Wisconsin—Stevens Point and part of the SCAR "Celebration of Women in Antarctic Research." Dr. Baeseman received an M.Sc. in civil engineering from the University of Minnesota and a Ph.D. in civil engineering with an environmental emphasis from the University of Colorado. Her first engagement with the National Academies was the Polar Research Board's International Polar Year planning in 2007, and she is a member of that board.

Scott Goetz, Ph.D., is Regents Professor at Northern Arizona University, with prior appointments at the Woods Hole Research Center, the University of Maryland, and the National Aeronautics and Space Administration (NASA) Goddard Space Flight Center. He has conducted satellite remote sensing research over the past 30+ years, served on the Intergovernmental Panel on Climate Change Fifth Assessment Report, United Nations programs on REDD+, the U.S. Global Change Research Program, and interagency programs on carbon cycle science, climate change, and terrestrial ecology. He is science lead of NASA's 10-year Arctic Boreal Vulnerability Experiment, deputy principal investigator (PI) of NASA's Global Ecosystem Dynamics Investigation (space station lidar mission), fellow of the American Geophysical Union (AGU), and editor in chief of *Environmental Research Ecology*. Dr.

Goetz received a Ph.D. in geographic sciences from the University of Maryland, College Park. He was a member of the National Academies committees for Research at Multiple Scales: A Vision for Continental Scale Biology and Decadal Survey for Earth Science and Applications from Space: Panel on Marine and Terrestrial Ecosystems and Natural Resource Management and planning committees for Opportunities to Use Remote Sensing in Understanding Permafrost and Ecosystems and Understanding Northern Latitude Vegetation Greening and Browning.

Jamin Greenbaum, Ph.D., is an assistant research geophysicist at the Institute of Geophysics and Planetary Physics, a division of the Scripps Institution of Oceanography at the University of California—San Diego. His research is focused on characterizing processes related to ice–ocean interactions and improving constraints on subglacial and continental shelf geology; his work uses aerogeophysical and oceanographic data and numerical models. He has led or participated in 16 expeditions to Antarctica and two to Greenland where he has operated airborne, shipborne, and ground-based instruments to understand the ice and surrounding ocean. He is a former Fulbright U.S. Scholar and recipient of the Antarctica Service Medal and NASA Operation Ice Bridge Instrument Team Group Achievement Award. Dr. Greenbaum received a B.S. and M.S. in aerospace engineering and a Ph.D. in geophysics from the University of Texas at Austin. He served as a member of the National Academies committee on Future Directions for Southern Ocean and Antarctic Nearshore and Coastal Research.

Dr. Greenbaum plays an advisory role as the manager of JKB Scientific LLC, providing consulting services on antibullying and antiharassment in remote field settings for Batelle related to its planned proposal to lead the next U.S. Antarctic Program Antarctic Support Contract. He was also an uncompensated board member of the Ocean Research Project from October 2014 to April 2020. He received free software from Seequent Limited for geophysical data analysis without restrictions on the conduct of the research or the publication of the results.

Michael Hartinger, Ph.D., is a research scientist at the Space Science Institute and an associate researcher at the University of California—Los Angeles. He studies space weather, particularly phenomena in plasma in the near-Earth environment that affect human life and technology. His work lies at the intersection of multiple geoscience and atmospheric science disciplines, and he frequently conducts interdisciplinary research on a range of Sun–Earth interactions. Dr. Hartinger is PI of an NSF-supported Antarctic magnetometer network and received several NASA research grants, including a citizen science project using sound and sonification techniques to study space weather. He has led several space weather community initiatives, such as a Geospace Environment Modeling workshop focus group

(2016–2021). Since 2017, he has also served on the AGU Space Physics and Aeronomy Nomination Task Force to increase the diversity of awards nominations. Dr. Hartinger received a Ph.D. in geophysics and space physics from the University of California—Los Angeles. He was a member of the National Academies committee for the Decadal Survey for Solar and Space Physics (Heliophysics) 2024–2033: Panel on the Physics of Magnetospheres and is on the Polar Research Board.

Cana Uluak Itchuaqiyaq, Ph.D., is an Iñupiaq scholar and a tribal citizen of the Noorvik Native Community in Northwest Alaska. She is an assistant professor of technical and scientific communication at Virginia Polytechnic Institute and State University. Her work focuses on Arctic research, data governance, and research methodologies that support community-led initiatives and scientific innovation. She has extensive experience developing research methods, partnerships, and programs that facilitate collaboration between Arctic communities and research institutions. She is the academic lead for Arctic community-led, transdisciplinary initiatives that focus on language documentation, natural language processing, data literacy, digital archiving, and increasing local research capacities. She serves as a colead for the Interagency Arctic Research and Policy Committee Participatory Research and Indigenous Leadership in Research collaborative team and International Arctic Science Committee Fourth International Conference on Arctic Research and Planning Research Priority Team 7 on Services, Infrastructure, Logistics, and Technology. Dr. Itchuaqiyaq has received numerous honors, including the 2023 and 2025 Conference on College Composition and Communication Best Article on Philosophy or Theory of Technical or Scientific Communication. Dr. Itchuaqiyaq holds a Ph.D. in technical communication and rhetoric from Utah State University, an M.A. in communication, media, and persuasion from Idaho State University, and an A.L.B. in environmental studies from Harvard Extension School.

Andrew J. Lloyd, Ph.D., is a Lamont Assistant Research Professor at Lamont-Doherty Earth Observatory, Columbia University. His research focuses on imaging Earth's seismological structure and developing new inverse methods that use Glacial Isostatic Adjustment (GIA) observations to directly image 3-D mantle viscosity or ice mass change. These efforts, largely focused in the polar regions, contribute to understanding regional geology, tectonics, and solid Earth–ice interactions and improve fundamental understanding of how GIA observations sense Earth structure and ice history. Dr. Lloyd has spent seven field seasons in the Antarctic installing, servicing, and demobilizing temporary broadband seismic stations and assisting with other geophysical deployments (e.g., GPS). He received a B.S. in mathematics and an M.S. in geosciences from the Pennsylvania State University and a Ph.D. in Earth and planetary sciences from Washington University in St. Louis.

Britney E. Schmidt, Ph.D., is an associate professor of astronomy and earth and atmospheric sciences at Cornell University and was an associate professor at Georgia Tech. She also serves on the board of directors of the Planetary Society. Dr. Schmidt is a polar and planetary scientist who works on understanding the interaction of climate with the cryosphere on Earth by developing and deploying a range of tools, including robotic platforms and in situ instruments. She and her team also use polar environments and ecosystems as analogs for other icy ocean worlds and test technologies on Earth for use in space. Dr. Schmidt was named the Blavatnik Awards National Laureate for Physical Sciences and Engineering in 2024, John F. Nye lecturer by AGU in 2023, and one of the 100 Most Influential People of 2023 by *Time Magazine*. She received a B.S. in physics from the University of Arizona and an M.Sc. and a Ph.D. in geophysics and space physics from the University of California—Los Angeles. She has served as a member of the National Academies committees on Astrobiology Strategy for the Search for Life and the 2020 Astrophysics Decadal Survey panel on Exoplanets, Solar System, and Astrobiology.